Yosr Kadri
Ons Haddad
Amina Ben Brahim

Staphylococcus aureus and antibiotic resistance: What's new?

Yosr Kadri
Ons Haddad
Amina Ben Brahim

Staphylococcus aureus and antibiotic resistance: What's new?

ScienciaScripts

Imprint

Cover image: www.ingimage.com

This book is a translation from the original published under ISBN 978-620-6-72452-0.

Publisher:
Sciencia Scripts
is a trademark of
Dodo Books Indian Ocean Ltd. and OmniScriptum S.R.L publishing group

120 High Road, East Finchley, London, N2 9ED, United Kingdom
Str. Armeneasca 28/1, office 1, Chisinau MD-2012, Republic of Moldova, Europe
Printed at: see last page
ISBN: 978-620-8-32989-1

Contents

1 INTRODUCTION

S. aureus is a Gram-positive grape-cluster bacterium belonging to *the Staphylococcaceae* family [1]. *S. aureus* is part of the cutaneous-mucosal commensal flora of human beings [2]. Nevertheless, it is the most virulent species in the genus, causing hospital-acquired and community-acquired infections worldwide. This germ is endowed with numerous virulence factors such as adhesins that enable colonisation of the host, the capsule, enzymes (coagulase and staphylokinase) and toxins (Hemolysin a, в, 5 and Panton Valentine Leucocidin or PVL) that protect the bacterium against the host's defences and promote the spread of infection [3,4]. These factors are implicated in a variety of pathologies of varying degrees of severity. In fact, this bacterium is one of the most frequent causes of pyogenes infections, implicated in almost 45% of skin and soft tissue infections. emeBacteremia, which represents the 2nd most frequent cause of infection at around 20.8%, is responsible for a high mortality rate of 20-40%. Pneumonia, which accounts for around 18.3% of cases, has a high mortality rate if caused by PVL-secreting *S. aureus* responsible for necrotizing pneumonitis, with a mortality rate of around 30-75%. Infections related to medical devices and urinary tract infections account for only 3% and 6% respectively [1,2,5-8].

Staphylococcal infections remain a major cause of mortality and morbidity worldwide, despite the development of antimicrobial agents. Dramatic changes in the sensitivity of *S. aureus* to antibiotics have been reported, particularly penicillin and betalactam antibiotics. These bacteria have become increasingly multi-resistant and have recently developed resistance to vancomycin, which has been used for several years to treat MRSA strains [3].

Methicillin-resistant Staphylococcus aureus (MRSA) is a major cause of community and hospital infections worldwide [9]. MRSA was first identified more than five decades ago. Since then, it has undergone epidemiological expansion and rapid change, causing a wide range of infections that can lead to sepsis and death. The first epidemic was described in the 1960s. The Centers for Disease Control and Prevention (CDC) has classified MRSA as a "serious threat" because it is associated with prolonged hospital stays and therefore an economic burden, the emergence of clones with reduced susceptibility to a wide range of antimicrobial agents and significant mortality. According to the WHO, resistance to methicillin (MRSA) increases the death rate of infected patients by 64%, and according to Canadian data, death rates linked to MRSA bacteria constitute 18.6% in 2021 [10-12].

Hundreds of thousands of deaths worldwide are due to antimicrobial resistance. It is predicted that, if the problem is not adequately addressed, more than 10 million annual deaths will be caused by antimicrobial-resistant microorganisms by 2050, surpassing cancer-related deaths. Comprehensive and coordinated surveillance and reporting is the cornerstone of efforts to control antimicrobial resistance [13].

Hence the interest of this work, which included all strains of *S, aureus* isolated in the microbiology laboratory of the Fattouma Bourguiba University Hospital in Monastir between 2012 and 2022, and which has the following objectives:

- To study the epidemiology of *S, aureus* isolates over the years of the study in the different departments and the different samples taken.
- To study the resistance of *S. aureus* strains to different antimicrobial agents.

2 MATERIALS AND METHODS

1. Equipment

1.1. Type, population and location of study

Our study is retrospective and descriptive, covering a period of 11 years (from January 2012 to 31 December 2022). It includes all bacterial strains of *Staphylococcus aureus* isolated from all bacteriological samples. These samples were taken from patients hospitalised and/or consulting at the Fattouma Bourguiba University Hospital in Monastir as well as from patients consulting at the 1ere line health structures in the Monastir region (Basic Health Centres and regional hospitals).

1.2. Fattouma Bourguiba University Hospital, Monastir

This hospital is a university hospital centre that participates in university and post-graduate training.

It comprises 23 departments as follows:

- 15 medical departments: Cardiology A and B, Endocrinology, Internal Medicine, Pneumology, Rheumatology, Infectious Diseases, Gastro-Hepato-Enterology, Nephrology, Neurology, Psychiatry, Dermatology, Paediatrics, Emergency, Occupational Medicine, Intensive Care.
- 8 surgical departments: General Surgery, Orthopaedic Surgery, Urology, Neurosurgery, Paediatric Surgery, Ophthalmology, Ear, Nose and Throat Surgery, Anaesthesia and Intensive Care.

The number of active beds is 874

The average number of hospital admissions between 2012 and 2022 was 3,661.

2. Methods

2.1. Data collection

The source of the data used in this study was the "Sirscan" software, a computer file used in the microbiology laboratory of the Fattouma Bourguiba University Hospital in Monastir for data entry and statistical analysis.

For each isolate, the data collected are :

- Patient's department of origin
- The nature of the bacteriological sample
- Sensitivity of this bacterial strain to the various antibiotics tested

Redundant strains were eliminated manually.

2.2. Type of sampling

A wide variety of samples are taken in the laboratory. They are either :

- Blood cultures
- ECBU (Urine Cyto-Bacteriological Examination)
- Pus samples

- Superficial: by swabbing pus from a superinfected wound or skin lesion.
- Deep: Abces, collection of pus

- Respiratory samples: Sputum Cyto-Bacteriological Examination (SCE), Broncho Alveolar Lavage (BAL), Tracheal Aspiration (TA), Protected Distal Sampling (PDP)
- Samples taken from equipment (catheter, probe, etc.)
- Puncture fluids: cerebrospinal fluid (CSF), pleural fluid, peritoneal fluid, ascites fluid, joint fluid
- Other samples such as genital samples, ophthalmological samples, etc.

2.3 ISOLATION AND IDENTIFICATION OF S. AUREUS

Bacterial identification of the strain involves a number of bacteriological characteristics of the germ, in particular morphological, cultural and biochemical characteristics.

2.3.1. MORPHOLOGICAL CHARACTERISTICS

- Direct examination in the fresh state shows cocci of around 0.5 to 1 iim in diameter, typically arranged in clusters or grape bunches, sometimes organised in diplococci or short chains.
- Direct examination after Gram staining shows bacteria that stain purple, known as Gram-positive bacteria **(Figure 1).**

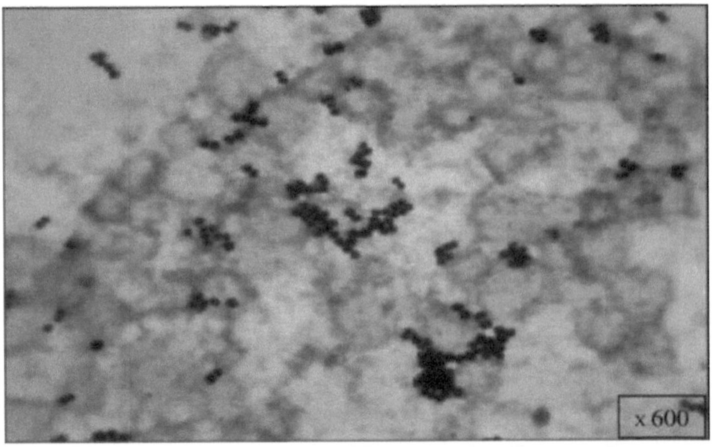

Figure 1: Appearance of S. aureus by Gram stain

2.3.2. Cultural characteristics

It is a non-demanding facultative aerobic-anaerobic germ that grows on ordinary gelose after 18 to 24 hours and tolerates temperature variations (24°-37°) as well as PH variations.

This bacterium gives the appearance of fairly large colonies, around 1 mm in diameter, which are round, regular, bulging, smooth and shiny (Smooth type). They are also creamy or yellow-pigmented.

Yellow-gold pigmentation can point to *S. aureus* **(Figure 2).**

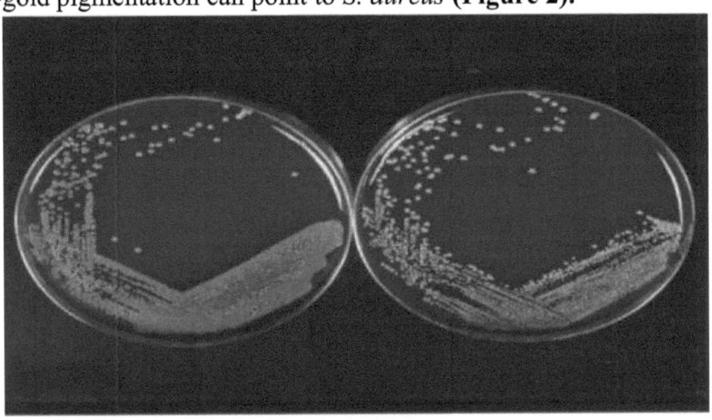

Figure 2: Appearance of S. aureus in culture

Polymicrobial samples are inoculated on selective media, in particular Chapman medium. The presence of mannitol and the high concentration of Nacl in this medium (7.5%) selects for halotolerant bacteria. The colonies of *S. aureus* on Chapman gelose show a colour change from red to yellow.

This change is due to the fermentation of mannitol and the acidification of the medium **(Figure 3)**.

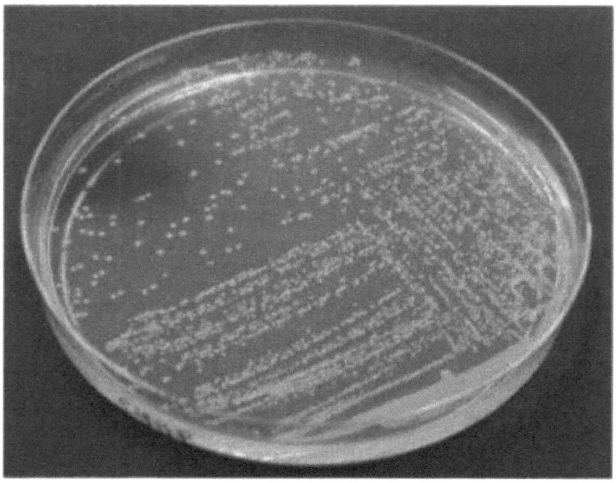

Figure 3: Culture of S. aureus on Chapman medium

2.3.3. BIOCHEMICAL CHARACTERISTICS

These characteristics help to identify and sometimes confirm the bacterial strain.

2.3.3.1 Gender identification

2.3.3.1.1. Catalase test

This test makes it possible to distinguish between the *Staphylcoccus* genus and the other Gram-positive cocci genera. It consists of demonstrating the production of the enzyme catalase by the bacteria tested in contact with hydrogen peroxide. A positive test results in the release of bubbles.

A positive catalase test points to the genus *Staphylococcus* **(figure 4)**.

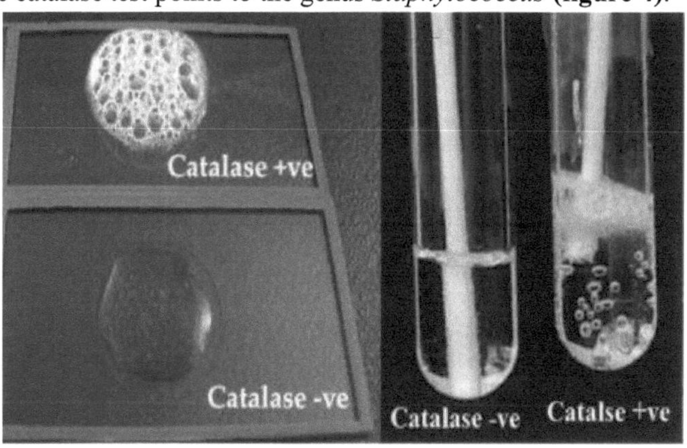

Figure 4: Catalase test

Catalase positive: Appearance of bubbles, the bacterium is said to be catalase (+)

Catalase negative: No bubbles appear, the bacteria are said to be catalase (-)

2.3.3.2 Species identification

2.3.3.2.1. Free coagulase test (Figure 5)

Coagulase or staphylocoagulase is an enzyme capable of coagulating blood plasma by converting fibrinogen into fibrin following the formation of staphylothrombin.

The coagulase test consists of mixing 0.5 ml of rabbit serum with 0.5 ml of a dense suspension of the bacteria and incubating for 24 hours at 37°C. A positive test is indicated by the formation of clots. However, observation of the reaction every 2 hours is necessary because of the risk of redissolution of the clot formed by fibrinolysis.

This test is used to differentiate *S. aureus* from other Coagulase Negative species (SCN).

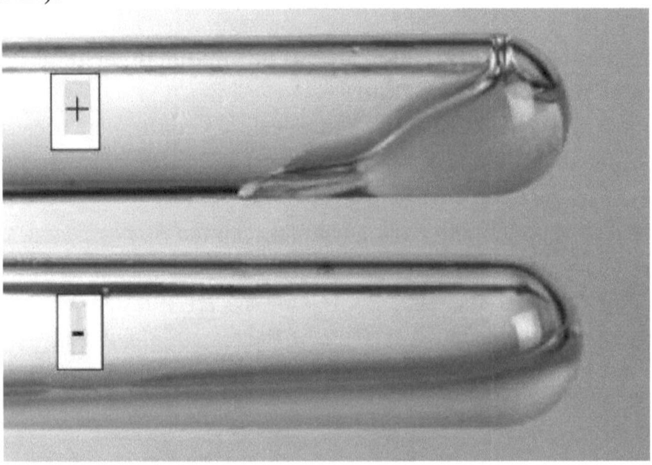

Figure 5: Free coagulase test

2.3.3.2.2. Desoxyribonuclease (DNase) test

This test is used to determine the ability of the pathogen to hydrolyse DNA and use it as a source of carbon and energy for growth.

This test is performed as follows:

- Inoculate a streak of the isolated strain onto a gelose containing DNA.
- Flood the gelose with 1:10 diluted HCl (1M) after 18 hours incubation.

A positive reaction results in the formation of a clear halo which doubles the Staphylococcus streak in 5 to 10 minutes **(figure 6)**.

This test is used to differentiate *S. aureus* from other species of Staphylococcus.

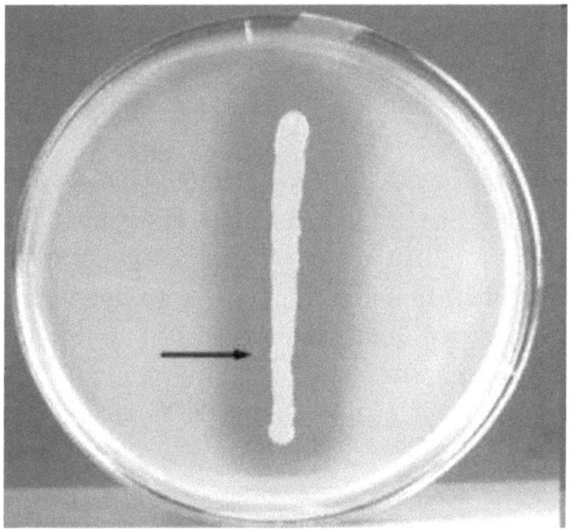

Figure 6: Detection of DNase

2.3.3.2.3. Culture on Chapman medium

The presence of mannitol and phenol red indicates the mannitol nature of the bacteria. Mannitol fermentation produces an acidic compound that turns the phenol red in the gelose yellow. This characteristic favours *the aureus* species **(Figure 7)**.

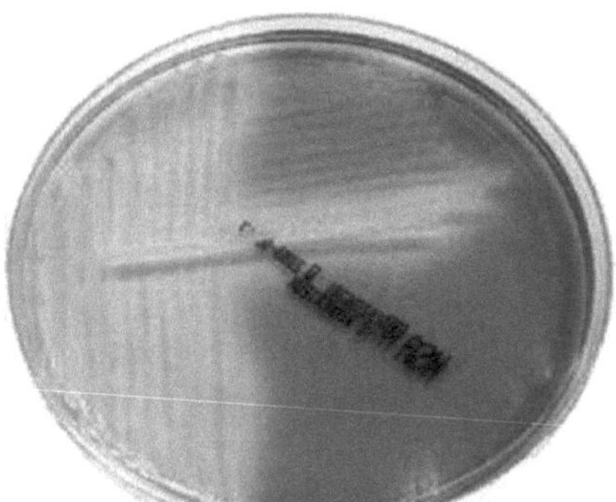

Figure 7: Fermentation of mannitol Turning the surrounding medium yellow

2.3.3.2.4. The VITEK® PLC GP card

This is a biochemical identification by the VITEK automated system using specific Gram-positive bacterial identification cards called GP cards.

2.4. Performing an antibiogram

In addition to bacterial identification, an antibiotic susceptibility test should be carried out, given the frequency of antibiotic-resistant strains and to demonstrate acquired antibiotic resistance.

Antibiotic susceptibility testing must be carried out in accordance with the recommendations of the Comite de l'Antibiogramme de la Societe Frangaise de Microbiologie (CA-SFM) and The European Committee on Antimicrobial Susceptibility Testing (EUCAST).

The technique used is the gelose diffusion method (gelose Mueller-Hinton). The inoculum is 0.5 McFarland with incubation in a normal atmosphere at a temperature of 35±2°C for 20±4 hours.

The determination of the Minimum Inhibitory Concentration (MIC) in certain situations is also carried out in accordance with CASFM-EUCAST recommendations.

The reference strain used for quality control is *Staphylococcus aureus* ATCC 29213.

2.4.1. ANTIBIOTICS TESTED

The antibiotics tested were :

penicillin G (1U), o\acillin(5Lg) and/or cefoxitin(30^g), kanamycin(30^g), tobramycin(10^g), gentamicin(10^g), tetracycline(30^g), chloramphenicol(30^g), erythromycin(15^g), lincomycin(15iig), pnstinamycin(15iig) and/or quinupristin-dalfopristin (15p.g), ofloxacin(5^g), norfloxacin(10^g), ciprofloxacin(5^g), le\4)llo\acin(5iig), lusiclic]iic acid(1(.)iig), trimethroprime+sulfonamides (1.25+23.75p.g), fosfomycin (50 ng), rifampicin (5p,g), vancomycin (5tig) and teicoplanin(30^g).

2.4.2. READING THE ANTIBIOGRAM

The EUCAST recommendations were followed for the first ten years of our study. Strains of intermediate susceptibility are classified as resistant.

Glycopeptides are an alternative to oxacillin in cases of resistance or intolerance. Resistance to these antibiotics is detected by determining MICs by microdilution in a liquid medium, which is the reference method.

The diffusion method (disc and strip gradient) cannot be used, as it does not allow differentiation between strains sensitive to glycopeptides and those with reduced sensitivity.

According to the CA-SFM, *S. aureus* strains with a MIC of vancomycin and/or teicoplanin > 1 mg/L by microdilution in liquid medium are rendered resistant.

2.5.STATISTICAL STUDY

The database and statistical analysis were carried out using Excel 2013 and SPSS (IBM SPSS Statistics 20).

The test used is the x-test^2 (Chi-square test).

The statistical significance level is p= 0.05

- If $p < 0.05$: the result is statistically significant
- If $p > 0.05$: the result is statistically insignificant
- NA: test Not Applicable.

3 RESULTS

1. DISTRIBUTION OF S. AUREUS STRAINS BY YEAR OF STUDY

During the study period, 34810 bacteria were isolated in the microbiology laboratory of the Fattouma Bourguiba University Hospital in Monastir. Among these bacteria, 4541 were *S. aureus* strains with an average of 412 strains per year, i.e. a prevalence of 13%. The prevalence ranged from 10.7% to 19%. The highest value was recorded in 2022 (19%, n=491), while the lowest was observed in 2020 (10.7%, n=291) **(Table I)**.

Table I: Prevalence of S. aureus strains by year of study of study

Year	Number of *S. aureus* strains	Total number of isolates	Prevalence (%*)
2012	479	4124	11,6
2013	432	3734	11,6
2014	452	3553	12,7
2015	383	2755	13,9
2016	370	2683	13,8
2017	413	3147	13,1
2018	441	3252	13,6
2019	471	3682	12,8
2020	291	2712	**10,7**
2021	318	2584	12,3
2022	491	2584	**19**
Total	4541	34810	13

***: Ratio of *S. aureus* strains isolated during the year of study to the total number of bacteria isolated during the same year.**

2. DISTRIBUTION OF S. AUREUS ISOLATES BY DEPARTMENT

The distribution of *S. aureus* strains differed between departments. This bacterium was isolated mainly from patients hospitalised on surgical wards (n=1851, 40.8%), mainly on orthopaedic wards (n=799, 17.6%). On the medical wards, 1265 strains of *S. aureus* were isolated, a rate of 27.9%. Among these departments, the nephrology department ranked first (n=336, i.e. 7.4%), followed by the dermatology department (n=239, i.e. 5.3%). On the other hand, the highest prevalence was found in medical departments (18%), in particular dermatology (39.6%), followed by surgery (11.9%) and health facilities (10.4%) **(Table II)**.

Table II: Distribution of S. aureus isolates by department

Services	Workforce	Percentage (%*)	Number of positive samples	Prevalence (%**)
Surgical departments	**1851**	**40,8**	**15577**	**11,9**
Orthopaedics	799	17,6	2375	33,6

Other surgical services [1]	1052	23,2	13202	8
Medical services	**1265**	**27,9**	**7008**	**18**
Nephrology	336	7,4	2363	14,2
Dermatology	239	5,3	603	39,6
Other medical services [2]	690	15,2	4042	17,1
Intensive care units	**527**	**11,6**	**7480**	**7**
Anaesthesia and intensive care	379	8,3	4892	7,7
Medical resuscitation	148	3,3	2588	5,7
Health structures[3]	**457**	**10**	**4372**	**10,4**
Paediatrics department	**289**	**6,4**	**4064**	**7,1**
Emergency department	**152**	**3,3**	**4892**	**3,1**
Total	4541	100	43393	10,5

***: Ratio of the number of *S. aureus* strains isolated in the department to the total number of *S. aureus* strains.**

****: Ratio of the number of *S. aureus* strains isolated in the department to the total number of positive samples in the department concerned.**

[1] : Paediatric surgery, ENT, General surgery, Gynecology, Urology, Ophthalmology, Neurosurgery.

[2] Cardiology, Neonatology, Rheumatology, Pneumology, Endocrinology, Gastrology, Psychiatry, Internal Medicine, Neurology, Occupational Medicine, Infectious Diseases.

[3] : Local dispensaries and regional hospitals.

3. TRENDS BY YEAR IN S. AUREUS STRAINS ISOLATED FROM VARIOUS SERVICES

During the study period, the frequency of isolation of *S. aureus* strains in surgical departments ranged from 32.2% in 2017 (n=133) to 48.1% in 2013 (n=208). The percentages for medical departments ranged from 22.2% in 2022 (n=109) to 32.4% in 2012 (n= 155). In addition, the frequencies of *S. aureus* strains fluctuated between 2.2% in 2014 (n=10) and 5.8% in 2020 (n=17) in the emergency department and between 4.2% in 2014 (n=19) and 2019 (n=20) and 9.7% in 2021 (n=31) in the paediatric department. For intensive care units, the percentage varies between 5.6% in 2013 (n=24) and 17.1% in 2022 (n=84) **(Figure 8)**.

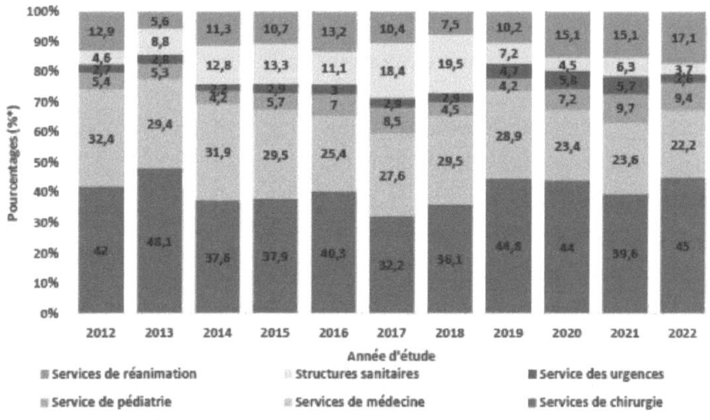

Figure 8: Frequency of *5. aureus* strains by year and department

* Ratio of *S. aureus* strains isolated in the department during the study year to the total number of *S. aureus* strains isolated during the same year.

surgical departments: orthopaedic surgery, paediatric surgery, oRL, general surgery, gynaecology, urology, ophthalmology, neurosurgery.

medical departments: Nephrology, Dermatology, Cardiology, Neonatology, Rheumatology, Pneumology, Endocrinology, Gastrology, Psychiatry, Internal Medicine, Neurology, Occupational Medicine, Infectious Diseases.

Health structures: Local dispensaries and regional hospitals.

A breakdown of the *S. aureus* isolates from the departments most concerned showed that in the orthopaedics department, the rate of isolated strains was stable, ranging from 12.3% in 2017 (n=51) to 24% in 2013 (n=102). In the nephrology department, the rate of collected strains reached 11% in 2018 and 2019, and was low in the other years. From 2018, the rate of *S, aureus* isolates decreased in the dermatology department. It fell from 7% to around 4% **(Figure 9)**.

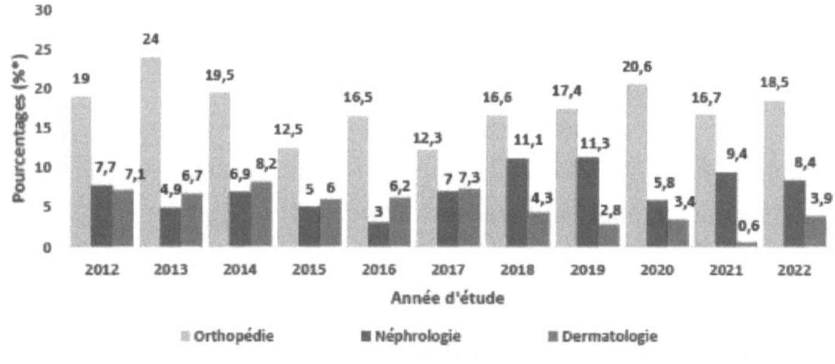

Figure 9: Percentage of *5. aureus* strains per year in orthopaedics, nephrology and dermatology departments

* Ratio of strains of *5. aureus* isolated in the department during the study year to the total number of strains *of 5.*

***aureus* isolated in the same year.**

4. DISTRIBUTION OF S. AUREUS ISOLATES PER SAMPLE

4.1. NUMBER, FREQUENCY AND PREVALENCE OF *5. AUREU5* ISOLATES BY DRAFT

Approximately half of the *S, aureus* strains were isolated from superficial pus samples (n = 1916, or 42.2%), blood cultures ranked 2eme (n = 825, or 18.2%) followed by deep pus samples (n = 762, or 16.8%). The highest prevalence was also observed for superficial pus samples (23.7%), followed by deep pus samples (17.7%) and blood cultures (16.7%) **(Table III).**

Table III: Number, frequency and prevalence of S. aureus isolates per collection

Type of levy	Number of samples positive for *S. aureus*	Percentage (%*)	Total number of positive samples	Prevalence (%**)
Superficial pus samples	1916	**42,2**	8080	**23,7**
Blood culture	825	**18,2**	4945	**16,7**
Deep pus sampling	762	**16,8**	4297	**17,7**
Respiratory sampling[1]	439	9,7	5066	8,7
ECBU[2]	199	4,4	19643	1
Puncture[3]	166	3,6	1263	13,1
Equipment[4]	144	3,1	2170	6,6
Other[5]	90	2	1438	6,2
Total	4541	100	46902	9,7

***: Ratio of the number of *S. aureus* positive samples to the total number of isolated *S. aureus* strains. **: Ratio of the number of *S. aureus* positive samples to the total number of positive samples.**

[1] Cytobacteriological examination of sputum, Tracheal aspiration, Bronchoalveolar lavage, Protected distal swab.

[2] Cytobacteriological examination of urine.

[3] Spinal fluid puncture, ascites puncture, joint puncture.

[4] Catheter, Drain/Redon.

[5] Sphere ENT, Genital sampling, Ophthalmological sampling, Biopsy, Bones.

4.2. EVOLUTION BY YEAR OF THE STRAINS ISOLATED IN THE DIFFERENT BACTERIOLOGICAL SAMPLES

During the study period, the rate of isolation of *S. aureus* from superficial pus samples ranged from 49.3% to 62.7%. The rate decreased in 2017, 2020, 2021 and 2022. For deep pus samples, the frequency ranged from 4.2% in 2014 (n=19) to 33.9% in 2017 (n=140). The percentages of blood cultures and ECBU positive for *S, aureus* varied between 12.7% in 2016 (n=47), 24.5% in 2021 (n=78) and 1.6% in 2016 (n=13) ,10% in 2020 (n=29), respectively **(Table IV).**

Table IV: Trends in *S. aureus* strains isolated from bacteriological samples by year

bacteriological samples

Prelevement	More	Deeper	ECBU[1]	Blood culture	Respiratory	Equipment[3]	Other[4]	Puncture[5]

(%*) "Ann6e \^^ˢ	superficial				sampling[2]			
2012	**49,3**	11,9	3,1	15,7	9	3,1	3,1	4,8
2013	**62,7**	6,2	3	16,4	3,7	2,5	2,5	3
2014	60,4	**4,2**	4,4	15,7	6,2	3,5	1,5	4,1
2015	50,7	11,7	3,7	16,4	8,4	1,8	2,9	4,4
2016	51,9	17	**1,6**	**12,7**	9,7	3	1,1	3
2017	**24,9**	**33,9**	3,9	18,2	8,2	4,6	1	5,3
2018	44	14,1	2,5	23,8	8,4	2,9	2	2,3
2019	39	15,5	7,9	20,8	8,1	3	2,5	3,2
2020	**23,7**	25,1	**10**	15,1	12,7	5,2	4,8	3,4
2021	**21,4**	25,5	6,9	**24,5**	14,2	3,1	0,3	4,1
2022	**26,9**	24,8	3,3	20	**18,9**	2,6	0,4	3,1

***: Ratio of *S. aureus-positive* samples collected during the study year to the total number of S. aureus strains isolated during the same year.**

1 Cytobacteriological examination of urine.

2 Cytobacteriological examination of sputum, Tracheal aspiration, Bronchoalveolar lavage, Protected distal swab.

3 Catheter, Drain/Redon.

4 Sphere ENT, Genital sampling, Ophthalmological sampling, Biopsy, Bones.

5 Spinal fluid puncture, ascites puncture, joint puncture.

5. DISTRIBUTION OF S. AUREUS STRAINS BY DEPARTMENT AND BY DRAFT

The distribution of *S, aureus* strains per sample differed between departments. *S, aureus-positive* superficial and deep pus samples were mainly isolated from surgical wards, with frequencies of 50.3% (n=964) and 70.4% (n=537), respectively. On the medical wards, *S. aureus-positive* blood cultures predominated (n=434, 52.6%). Respiratory samples positive for *S, aureus* were more common in intensive care units (n=317, 72.2%) **(Table V)**.

Table V: Distribution of *S. aureus* strains by department and by collection

Levy (%*) Service	More superficial	Deeper	Blood culture	ECBU[1]	Respiratory sampling[2]	Equipment[3]	Puncture[4]	Other[5]
Surgical departments	**50,3**	**70,4**	18,8	29,7	1,6	18,1	24,8	**69**
Medical services	26,6	12,5	**52,6**	20,6	8,7	42,4	46,4	10
Emergency department	0,8	1,6	4,6	18,6	4,1	9	10,8	1,1
Paediatrics department	5,6	5,4	9	0,5	9,3	4,2	9	4,4
Intensive care units	3	2,4	10,8	4	**72,2**	20,8	3,6	2,2
Health facilities	13,7	7,7	4,2	26,6	4,1	5,5	5,4	13,3

***: Ratio of *S. aureus* strains isolated from a sample in a given department to the total number of *S. aureus* strains isolated from the same sample.**

1 Cytobacteriological examination of urine.

2 Cytobacteriological examination of sputum, Tracheal aspiration, Bronchoalveolar lavage, Protected distal swab.

3 Catheter, Drain/Redon.

4 Spinal fluid puncture, ascites puncture, joint puncture.

5 Sphere ENT, Genital swab, Ophthalmic swab, Biopsy, Bone.

*** Orthopaedics, Paediatric surgery, ENT, General surgery, Gynecology, Urology, Ophthalmology, Neurosurgery.**

* *: Nephrology, Dermatology, Cardiology, Neonatology, Rheumatology, Pneumology, Endocrinology, Gastrology, Psychiatry, Internal Medicine, Neurology, Occupational Medicine, Infectious Diseases.
* : **Local dispensaries and regional hospitals.

6. TRENDS BY YEAR AND BY DEPARTMENT IN STRAINS *of S. AUREUS* COLLECTED IN DIFFERENT SAMPLES

In the surgical wards, there was a consistent predominance of pus collections over all years of **the study (Figure 10).**

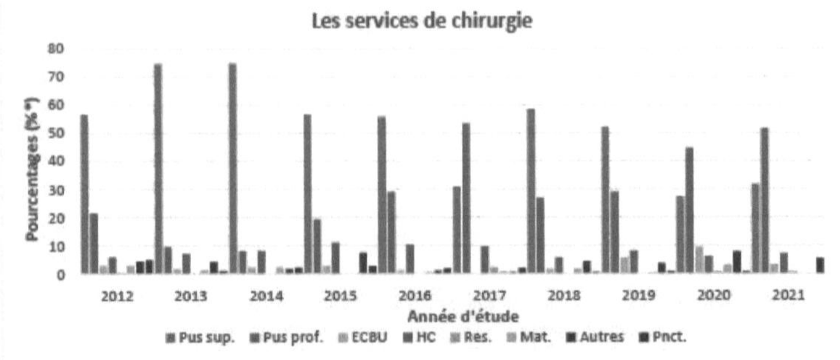

Figure 10: Trends in strains *of 5. aureus* collected in the various surgical departments

* **Ratio of *S. aureus* strains isolated in a given sample, within the department concerned, during the study year to the total number of isolates in the department.**

Over the eleven years of the study, a predominance of *S. aureus* isolates was observed in superficial pus samples taken from medical wards between 2012 and 2016. From 2017 onwards, strains were mainly collected from blood cultures **(Figure 11).**

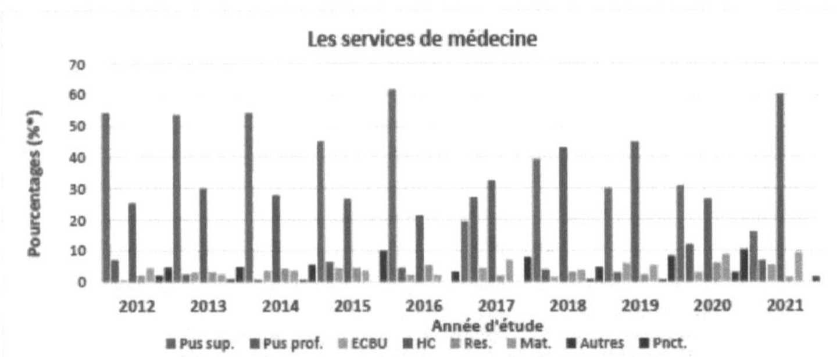

Figure 11: Trends in *S. aureus* strains collected in the various samples taken from medical departments

In intensive care units, almost all *S, aureus* isolates were collected from respiratory samples. In fact, the frequency of *S, aureus* in this type of sample has increased over the years, reaching 81% (n=84) in 2022 **(Figure 12).**

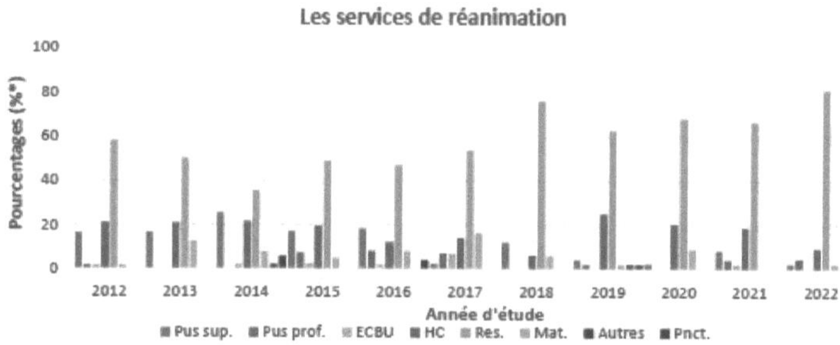

Figure 12: Trends in *S. aureus* strains collected in various samples from intensive care units

* **Ratio of *S. aureus* strains isolated in a given sample, within the department concerned, during the study year to the total number of isolates in the department.**

In the paediatric department, the majority of *S. aureus* strains were isolated from pus samples and blood cultures between 2012 and 2020. On the other hand, respiratory samples have increased and are at the top of the list in 2021 and 2022 **(Figure 13).**

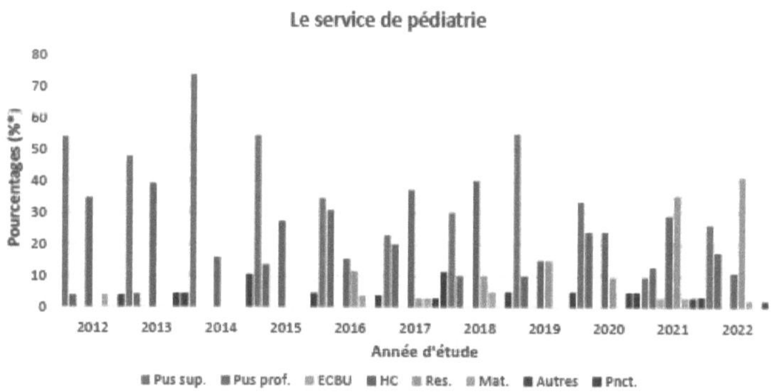

Figure 13: Trends in strains *of 5. aureus* collected in different samples from the paediatric ward

* **Ratio of *S. aureus* strains isolated in a given sample, within the department concerned, during the study year to the total number of isolates in the department.**

In the emergency department, the frequency of *S, aureus-positive* samples differed over the study years. In fact, *S. aureus-positive* ECBUs were in the majority in 2012 (30.8%, n=4), 2019 (50%, n=11) and 2020 (52.9%, n=9). The rate of isolates in blood cultures was highest in 2014 (30%, n=3), 2017 (33.3%, n=4), 2021 (33.3%, n=6) and 2022 (46.2%, n=6). For the rest of the years, *S. aureus-positive* pus samples predominated **(Figure 14).**

Figure 14: Trends in *S. aureus* strains collected in the various samples taken in the emergency department

At health facilities, superficial pus samples predominated between 2012 and 2019 and in 2022. However, ECBU samples were in the majority in 2020 and 2021 **(Figure 15).**

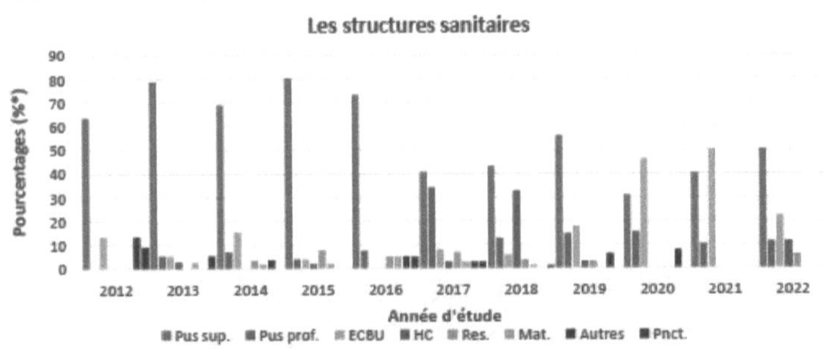

Figure 15: Trends in *S. aureus* strains collected in various samples taken at health facilities

*** Ratio of *S. aureus* strains isolated in a given sample, within the department concerned, during the study year to the total number of isolates in the department.**

7. STUDY OF THE RESISTANCE OF S. AUREUS STRAINS TO ANTIBIOTICS

7.1.REsISTANcE TAuX OF s. IsOLATs AuREus AuX ANTIBIOTIQuEs

The rate of resistance of *S. aureus* strains to penicillin G during the study period was very high, at around 96.4% (n=4296). The rate of cefoxitin resistance during this period was 18.6% (n=828) **(Figure 16)**.

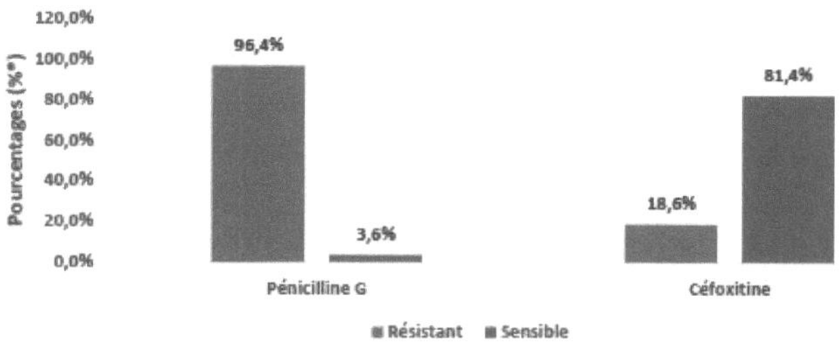

Figure 16: Resistance of *S. aureus* strains to beta-lactam antibiotics

***: Ratio of antibiotic-resistant *S. aureus* strains to the total number of strains tested.**

Resistance of isolates to kanamycin, tobramycin and gentamicin was 28% (n=1003), 11.5% (n=420) and 7.4% (n=214), respectively. Kanamycin responded to amikacin. As for macrolides, the resistance rate of our *S. aureus* strains to erythromycin was 13.8% (n=609). Resistance to lincomycin was 5.3% (n=187) and no resistance to pristinamycin was noted. Resistance to fluoroquinolones was 13.3% (n=334) to norfloxacin, 12.2% (n=412) to ofloxacin, 14.6% (n=130) to levofloxacin and 14.5% (n=130) to ciprofloxacin. The rates of resistance of *S. aureus* strains to vancomycin and teicoplanin were 1.2% (n=24) and 0.9% (n=21) respectively. Resistance to other families of antibiotics was variable, with the highest percentage for tetracycline (28.9%, n=363) followed by fusidic acid (27,1%, n=1136), chloramphenicol (1.9%, n=45), rifampicin (7.4%, n=279), the trimethoprim-sulphamethoxazole combination (5.9%, n=182) and fosfomycin (5.6%, n=142) **(Table VI).**

Table VI: Overall rate of resistance of *S. aureus* to other antibiotics

Antibiotic families	Antibiotics (number of strains tested)	Number of resistant strains	Resistance rate (%*)
	Kanamycin (n=3581)	1003	28
Aminosides	**Tobramycin (n=3640)**	420	11,5
	Gentamicin (n=2908)	214	7,4
	Erythromycin (n=4419)	609	13,8
Macrolides	**Lincosamycin (n=3558)**	187	5,3
	Pristinamycin (n=4265)	0	0
	Norfloxacin (n=2503)	334	13,3
Fluoroquinolones	**Ofloxacin (n=3381)**	412	12,2
	Levofloxacin (n=890)	130	14,6
	Ciprofloxacin (n=896)	130	14,5
Glycopeptides	**Vancomycin (n=1969)**	24	1,2
	Teicoplanin (n=2374)	21	0,9
	Tetracycline (n=1256)	363	28,9
	Chloramphenicol (n=2426)	45	1,9

Other	Fusidic acid (n=4199)	1136	27,1
	Trimethoprimesulfamethoxazole (n=3109)	182	5,9
	Rifampicin (n=3761)	279	7,4
	Fosfomycin (n=2542)	142	5,6

***: Ratio of antibiotic-resistant *S. aureus* strains to the total number of strains tested.**

7.2.EVOLUTION OF ANTIBIOTIC RESISTANCE IN ISOLATES OVER TIME YEARS

Over the years, there has been no change in penicillin G resistance. The rates were high whatever the year, the sample and the department. The revolution in methicillin resistance and its distribution in the different departments will be detailed in a separate chapter.

Resistance to kanamycin ranged from 24.3% (n=91) to 32.8% (n=127) with a higher figure in 2020 (42.9%, n=39) (p=0.019). Gentamicin and tobramycin resistance was stable over the years (NA; p<0.001, respectively) **(Figure 17).**

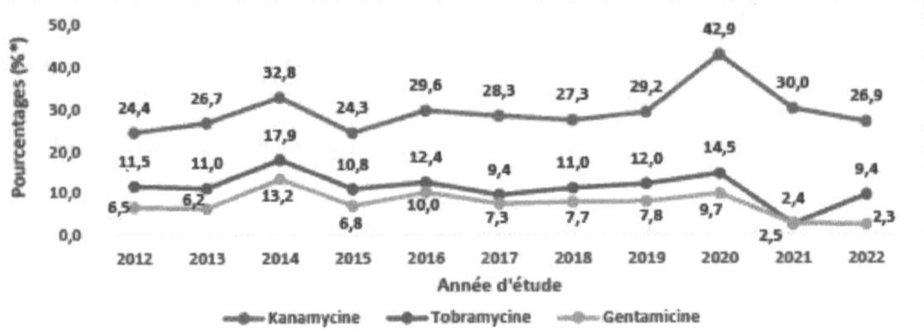

Figure 17: Evolution of resistance of isolates to aminoglycosides

The rate of resistance to erythromycin a ranged from 8.1% (n=33) to 19.6% (n=86) (p<0.001). Resistance to lincomycin a ranged from 1.9% (n=6) to 9.6% (n=42) (NA). Nevertheless, no resistance to pristinamycin was noted during this period **(Figure 18).**

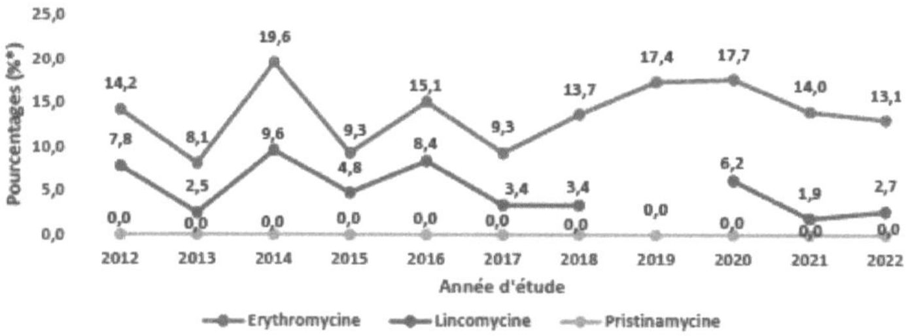

Figure 18: Evolution of macrolide resistance in isolates

The rates of resistance of *S. aureus* isolates to fluoroquinolones have evolved in parallel and have varied between 6.7% and 17.7%. The highest rate of resistance to this family of antibiotics was recorded in 2014 **(Table VII).** This result was statistically significant only for ofloxacin (p=0.002).

Table VII: Trends in resistance of isolates to fluoroquinolones

	2012	2013	2014	2015	2016	2017	2018	2019	2020	2021	2022
Norfloxacin (%)	14,1	NT	17,7	11,7	14,3	NT	12,7	14,8	9,8	8,7	13,8
Ofloxacin (%)	14,1	NT	17,7	11,5	13,4	6,7	11,4	14,5	8,3	9,8	12,8
Levofloxacin (%)	14,1	NT	17,7	11,7	14,3	NT	NT	NT	NT	NT	NT
Ciprofloxacin (%)	14,1	NT	17,7	11,7	14,3	NT	NT	NT	NT	0	0

NT: Non Teste.

As for glycopeptides, no resistance was detected in 2012, 2017 and 2018. In 2014 and 2016, a single resistant strain was isolated from a blood culture in the intensive care unit. In 2015, the resistant strain was detected in a deep pus sample in health facilities.

In 2019, a single strain was found to be resistant to vancomycin and was also isolated in intensive care units from a superficial pus sample.

As for teicoplanin, 3 isolates from superficial pus samples (2 from the intensive care unit and 1 from the ENT department) were resistant.

In 2020, no vancomycin-resistant strains were detected and 8.2% of isolates tested were resistant to teicoplanin (n=4). Half were isolated from surgical wards (in a deep pus sample and a respiratory sample), one strain was isolated from a material sample in the intensive care unit and the other was isolated from a superficial pus sample in the rheumatology department.

In 2021, vancomycin resistance increased to a peak of 16.9% (n=14). Of these strains, six were isolated from surgical departments, four from medical departments, two from intensive care units and the other two from emergency departments and health facilities. This resistance was detected in 6 blood

cultures, 4 superficial pus samples, two ECBU and two punctures. Resistance to teicoplanin was 8.6% (n=6). These isolates were detected in the nephrology department (n=1), orthopaedics department (n=2), intensive care unit (n=2) and emergency department (n=1). Half were superficial pus samples and half were blood cultures.
Resistance to vancomycin in 2022 was 6.3% (n=6). Two strains were isolated from medical wards, two from intensive care units and the other two were detected in surgical wards and health facilities (3 ECBU, 2 pus samples, 1 respiratory sample). As for teicoplanin, 5.3% of isolates were resistant (n=5). These strains were isolated from surgery (n=2), endocrinology (n=1), intensive care (n=1) and health facilities (n=1) (2 deep pus samples, 2 ECBU and 1 respiratory sample) **(Figure 19).** The statistical test is not applicable (NA) for these 2 antibiotics because there are cells with a theoretical number of cells of less than 5.

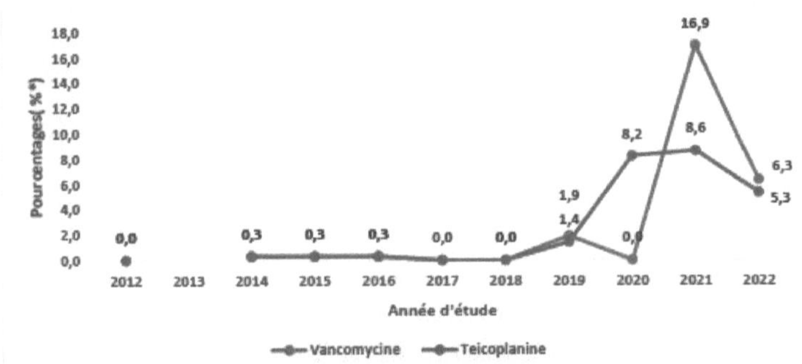

Figure 19: Evolution of glycopeptide resistance in isolates

During our study period, there was a significant fluctuation in the antibiotic resistance rates of *S. aureus* isolates. Between 2012 and 2013, there was a decrease in resistance to fusidic acid and rifampicin. From 2014 onwards, there was an increase in the rate of resistance to fosfomycin until 2016, then a decrease was noted between 2017 and 2022. There was also a steady rise in the rate of resistance of *S. aureus* strains to fusidic acid, which reached 40.6% (n=191) in 2022. Nevertheless, the evolution of rifampicin resistance was relatively stable **(Figure 20).** These results were statistically significant for fusidic acid ($p<0.001$) and fosfomycin ($p<0.001$). However, the statistical test was not applicable for rifampicin.

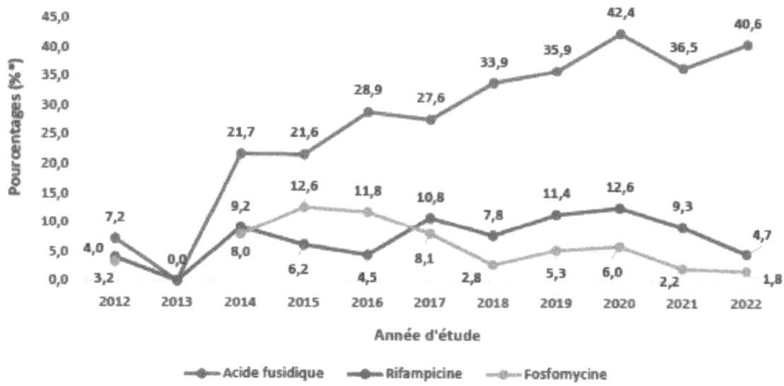

Figure 20: Evolution of isolate resistance to rifampicin, fosfomycin and fusidic acid

As for tetracycline, there was an increase in the rate of resistance from 2014 to 2016, followed by a decrease in the percentage of *S.aureus* strains resistant to this antibiotic, reaching 1.4% (n=1) in 2018. Of the two strains tested in 2019 and 2020, one was resistant to this antibiotic. The trend in resistance to chloramphenicol and trimethoprimesulfamethoxazole was relatively stable **(Figure 21).** These results were statistically significant for the trimethoprim-sulfamethoxazole combination (p=0.011). However, the statistical test is not applicable for the rest of the antibiotics because there are cells with a theoretical size of less than 5.

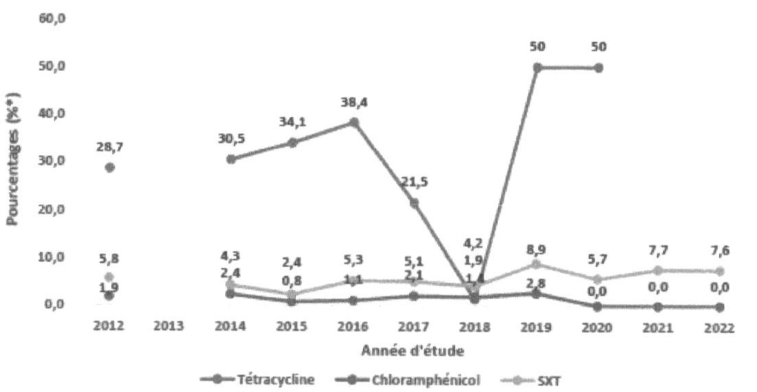

Figure 21: Evolution of resistance in isolates to tetracycline, chloramphenicol and trimethoprim-sulfamethoxazole combination

***: Ratio of antibiotic-resistant *S. aureus* strains in the study year to the total number of strains tested in the same year.**

7.3 ANTIBIOTIC RESISTANCE OF S. AUREUS STRAINS IN DIFFERENT DEPARTMENTS

The resistance of *S. aureus* isolates to the different families of antibiotics varied according to the department. In fact, the highest rates of resistance were observed mainly in intensive care units, the emergency department and medical units **(Table VIII).** This difference in the percentage of resistance between departments was statistically significant for all the antibiotics tested ($p<0.05$) with the exception of tetracycline ($p=0.630$), fusidic acid ($p=0.165$) and the trimethoprim-sulfamethoxazole combination ($p=0.067$).

Table VIII: Antibiotic resistance rates of S. aureus according to services

Levy \(%)* Antibiotic	Surgical departments[1]	Medical services[2]	Intensive care unit	Health structures[3]	Paediatrics department	Emergency department	p-value
Penicillin G	**97,3**	**95,2**	**95,8**	**96**	**97,9**	**94,7**	**0,021**
Cefoxitin	**18,8**	**16,3**	**28,8**	**17,3**	**10,9**	**20,3**	**0,000**
Kanamycin	**27,3**	**28,9**	**37,1**	**24,6**	**20**	**23,4**	**0,000**
Tobramycin	**9,8**	**11,6**	**25,5**	**7,1**	**5,5**	**10,7**	**0,000**
Gentamicin	**6,2**	**6,7**	**20,3**	**2,2**	**1,1**	**7,5**	**0,000**
Erythromycin	**12,8**	**12,9**	**17,6**	**14,7**	**16,5**	**11,6**	**0,045**
Lincomycin	**4,5**	**3,7**	**12,2**	**4,5**	**5**	**4,5**	**0 ,000**
Pristinamycin	0	0	0	0	0	0	
Norfloxacin	**11,2**	**14**	**23,5**	**10,5**	**4,8**	**20,9**	**0,000**
Ofloxacin	**10,1**	**13,6**	**21,7**	**9,3**	**4,4**	**15,1**	**0,000**
Levofloxacin	10 ,1	14,7	**34,2**	11,9	1,9	22,7	NA
Ciprofloxacin	10	14,7	**33,6**	11,9	1,8	21,7	NA
Tetracycline	28,4	**31,2**	31,1	27,3	23	22,6	0,630
Chloramphenico l	1,3	**3**	1,9	1,8	0	1,2	NA
Fusidic acid	26,6	28,2	22,9	28,1	28	**32,6**	0,165
SXT†	5,7	6	**8,9**	4,5	3,4	4,5	0,067
Rifampicin	**6,5**	**7,5**	**11,6**	**7,3**	**3,8**	**11,1**	**0,001**
Fosfomycin	**4,7**	**5,9**	**12**	**3,5**	**1,7**	**3,3**	**0,000**
Vancomycin	1,1	1,1	**2,3**	1,3	0	1,9	NA
Teicoplanin	0,8	0,5	**2,7**	0,7	0	1,4	NA

- Trends in resistance to aminoglycosides in different departments

In the majority of departments, isolate resistance to kanamycin was stable over the study years, with the exception of medical departments, health facilities and intensive care units. A peak was noted in 2020 in medical wards and health facilities, with rates of 70.8% (n=17) and 66.7% (n=2) respectively. In intensive care, high rates were reported in 2014 (68.8%, n=33) and 2021 (50%, n=2). As for emergency departments, the rate was stable except in 2021. The only strain isolated was resistant to kanamycin **(Table IX).** Isolate resistance to tobramycin and gentamicin was stable in all departments except intensive care. Indeed, a

*** Ratio of *S. aureus* strains resistant to 1 antibiotic in the department to the total number of *S. aureus* strains tested for this antibiotic in the same department.**

† Trimethoprim-sulfamethoxazole.

[1] Orthopaedics, paediatric surgery, ENT, general surgery, gynaecology, urology, ophthalmology, neurosurgery.

[2] Nephrology, Dermatology, Infectious diseases, Cardiology, Neonatology, Rheumatology, Pneumology, Endocrinology, Gastrology, Psychiatry, Internal medicine, Neurology, Occupational medicine.

[3]: Local dispensaries and regional hospitals.

peak was observed for these 2 antibiotics in 2014 with a rate of 56.3% for tobramycin (n=27) and 50% for gentamicin (n=19) (**Table IX).**

Table IX: Trends in resistance of *S. aureus* isolates to aminoglycosides in different departments

Services \ (%*)	Services medicine[1]			Services intensive			Services care/emergency			Surgery[2]			Paediatrics			Health facilities[3]		
Year	KN	TB	GM	KN	TB	GM	KN	TB	GM	KN	TB	GM	KN	TB	GM	KN	TB	GM
2012	25	П,5	8,1	20,3	9,3	3,1	26,7	21,7	16,7	36	0	0	33,3	16,7	16,7	31,8	13,6	0
2013	28,4	6,9	3,9	25,1	12	6,5	37,5	20,8	20,8	20	15	5	22,2	11,1	11,1	27,3	9,1	0
2014	23,7	11	7,5	34,5	16,9	12,5	68,8	56,3	50	H,1	5,6	5,6	22,2	22,2	12,5	25	3,9	20,8
2015	23,1	13,2	4,8	23,6	4,6	4,8	35	40	38,5	22,7	0	0	36,4	0	0	18	4,3	0
2016	31,9	14,6	6,4	31,3	7,2	7,3	36,7	33,3	33,3	8,7	0	0	18,2	18,2	11,1	24,4	7,5	5,7
2017	33	8,1	7,1	30,3	11,8	8,9	26,3	20,6	19,4	25	3,4	0	0	0	0	26	4,8	1,9
2018	29,7	10,8	15,8	26,6	7,1	5,9	28,1	9,1	0	10	10	0	46,2	25	0	25,9	9,5	0
2019	28,4	11,4	7,6	31,9	11,8	8,7	40	26,1	16,1	18,2	11,1	0	15,4	0	0	12,5	3,2	0
2020	70,8	23,7	18,4	31,6	10,8	4,2	33,3	15,4	14,3	0	0	0	25	9,1	7,1	66,7	16,7	20
2021	0	0	0	27,3	1,7	2,1	50	1,3	7,9	0	0	0	100	10	7,1	0	0	0
2022	30,3	9,5	1,5	34,6	7	2,6	34,3	15	2,1	22,5	10,5	0	8,3	7,7	0	26,7	12,5	7,7

KN: Kanamycin; TB: Tobramycin; GM: Gentamicin.

***: Ratio of *S. aureus* strains resistant to the antibiotic in the department during the study year to the total number of *S. aureus* strains tested for this antibiotic in the department during the same year.**

4 Nephrology, Dermatology, Infectious diseases, Cardiology, Neonatology, Rheumatology, Pneumology, Endocrinology, Gastrology, Psychiatric, Internal medicine, Neurologic, Occupational medicine.

[2] **: Orthopedic, Paediatric surgery, ENT, General surgery, Gynecologic, Urology, Ophthalmology, Neurosurgery.**

[3] **'■ Local dispensaries and regional hospitals.**

- Trends in macrolide resistance in different departments

Resistance to erythromycin was stable in the majority of departments, ranging from 0% to 33.3% (n=4). The exception was intensive care units, which showed a peak in frequency in 2014 (43.8%, n=21). Lincomycin resistance was also stable in the majority of departments, ranging from 0% to 20% (n=6), with the exception of intensive care departments, where there were 2 frequency peaks in 2014 (35.4%, n=17) and 2016 (29.2%, n=14), and the paediatric department, which showed a significant increase in the rate in 2018 (33.3%, n=1) **(Table X).**

Table X: Trends in macrolide resistance of X *aureus* isolates in different departments

'''к Services	Medical services [1]	Surgical departments[2]	Services de resuscitation	Paediatrics department	Emergency department	Structures washrooms[‡]

[‡]Nephrology, Dermatologic, Infectious diseases, Cardiologic, Neonatology, Rheumatology, Pneumology, Endocrinology, Gastrology, Psychiatric, Internal medicine, Neurologic, Occupational medicine.

[2] **: Orthopedic, Paediatric surgery, ENT, Surgery, Gynecologic, Urology, Ophthalmology, Neurosurgery.**

[3] **: Local dispensaries and regional hospitals.**

(%*)												
	Eryth	**Linco**	**Eryth**	**Linco**	**Eryth**	**Linco**	**Eryth**	**Linco**	**Eryth**	**Linco**	**Eryth**	**Linco**
2012	16,1	9,4	9,9	5,2	16,7	13,3	7,7	3,8	33,3	16,7	30	5
2013	6,9	0,9	8	4	4,5	0	9,1	4,5	10	0	14,3	0
2014	15,5	4,9	18,2	8,5	43,8	35,4	21,1	10,5	0	11,1	16,1	1,8
2015	8,7	2,8	7,9	4,2	15	20	9,1	4,5	9,1	0	10	0
2016	12,8	6,4	11,8	4,9	32,7	29,2	8	5	9,1	0	17,1	4,9
2017	10,1	1	5,5	1,6	20,9	4,7	3	0	0	8,3	12,3	10,6
2018	17,1	0	13,3	8,3	18,2	0	15	33,3	7,7	0	8,2	0
2019	14,4	NT	21,4	NT	10,4	NT	15	NT	13,6	NT	18,2	NT
2020	13,8	5	20,6	10,5	9,5	0	30	0	6,3	0	30,8	7,7
2021	9,6	0	11,2	0	16,7	6,7	22,6	0	11,1	0	31,6	15,8
2022	13,9	1,8	11,4	1,8	6	1,2	32,6	10,9	23,1	7,7	5,6	0

Eryth: Erythromycin; Linco: Lincomycin.
NT: Non Teste

***: Ratio of strains of 5. *aureus* resistant to 1 antibiotic in the department during the study year to the total number of strains of 5. *aureus* tested for this antibiotic in the department concerned during the same year.**

7.4 RESISTANCE OF S. AUREUS TO ANTIBIOTICS ACCORDING TO THE NATURE OF THE sample

During the study period, there was a significant difference in the percentages of isolate resistance according to the nature of the sample. Material, punctures and ECBU were the most common sources of resistance to most antibiotics **(Table XI).** This variation was statistically significant for tobramycin (p=0.011), norfloxacin (p<0.001), ofloxacin (p<0.001) and fusidic acid (p<0.001).

Table XI: Antibiotic resistance rate of *S. aureus* by type of sample

Levy ^\ (%)§ Antibiotics^	More superficial	Deeper	ECBU[1]	Blood culture	Respiratory sampling[2]	Equipment[3]	Other [4]	Puncture [5]	Value -p
Penicillin G	96,7	**97,8**	92,7	95	97	97,1	93,3	96,3	NA
Cefoxitin	16,6	17,9	18,7	23,3	19,8	**24,5**	18,9	15,1	0,003
Kanamycin	29,2	24,4	**31,6**	26,1	28,7	30,2	32,9	26,9	0,350
Tobramycin	10,5	8,3	14,7	13,4	14,4	**17,4**	12,8	12,6	0,011
Gentamicin	6,1	4,9	7,1	10,6	10,4	**13,8**	7,7	5,9	NA**
Erythromycin	13,1	12,8	16,4	13,3	14,0	**20,0**	22,1	15,6	0,079
Lincomycin	5,9	2,4	6,5	4,8	6,8	**8,0**	11,1	2,2	NA***
Pristinamycin	0,0	0,0	0,0	0,0	0,0	0,0	0,0	0,0	
Norfloxacin	12,0	10,4	19,4	14,6	10,6	15,6	21,8	**26,4**	0,000
Ofloxacin	11,0	9,0	17,5	14,4	10,4	18,8	18,5	**19,3**	0,000
Levofloxacin	11,3	7,3	23,1	20,4	14,5	25,0	28,6	**25,6**	NA***
Ciprofloxacin	11,2	7,1	23,1	20,3	14,3	25,0	28,6	**25,6**	NA***

§ Percentage of *S. aureus* strains resistant to 1 antibiotic in relation to the total number of *S. aureus* strains tested for this antibiotic in the same sample.
**** Not applicable.**
[1] Cytobacteriological examination of urine.
[2] Cytobacteriological examination of sputum, Tracheal aspiration, Bronchoalveolar lavage, Protected distal swab.
[3] Catheter, Drain/Redon.
[4] Sphere ENT, Genital sampling, Ophthalmological sampling, Biopsy, Bones.
[5]Spinal fluid puncture, ascites puncture, joint puncture.

Tetracycline	**32,6**	25,0	28,3	28,4	23,8	22,2	15,8	24,1	0,195
Chloramphenicol	2,2	0,6	**2,9**	1,6	1,5	0,0	10,4	1,1	NA***
Fusidic acid	27,6	29,3	**34,3**	27,1	16,8	34,1	24,1	25,8	0,001
SXT††	4,9	6,1	6,1	7,2	4,0	**12,5**	10,2	6,3	NA***
Rifampicin	7,5	7,3	**10,5**	7,4	5,5	9,3	6,2	8,0	0,634
Fosfomycin	4,1	4,5	10,2	7,7	5,6	**13,3**	6,3	3,5	NA***
Vancomycin	0,7	0,6	5,5	**2,7**	0,5	0,0	0,0	2,2	NA***
Teicoplanin	0,7	1,1	2,2	1,2	0,9	**1,4**	0,0	0,0	NA***

8. METHICILLIN-RESISTANT *S. AUREUS* (MRSA)

8.1. DEVELOPMENT OF SARM OVER THE YEARS

Rates of methicillin resistance in *S. aureus* strains were relatively stable, ranging from 13.2% (n=57) to 24.2% (n=92) over the study period **(Figure 22)**. This result was statistically significant (p=0.002).

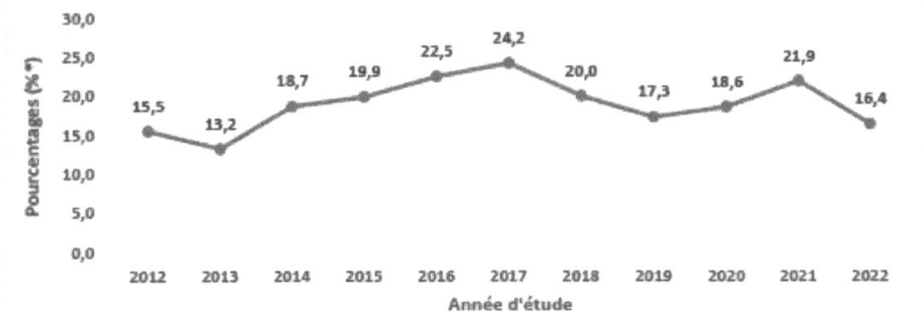

Figure 22: Changes in MRSA over the years studied

***: Ratio of MRSA strains during the study year to the total number of strains during that year.**

8.2. DISTRIBUTION OF MRSA TO DIFFERENT DEPARTMENTS

MRSA strains predominated in the intensive care unit, emergency department and surgery departments, with percentages of 28.8% (n=148), 20.3% (n=30) and 18.8% (n=341) respectively. The lowest rate of MRSA was noted in the paediatric department (10.9%, or n=31) **(Figure 23).** This variation in the percentages of *S. aureus* strain resistance was statistically significant (p<0.001).

†† Trimethoprim-sulfamethoxazole.

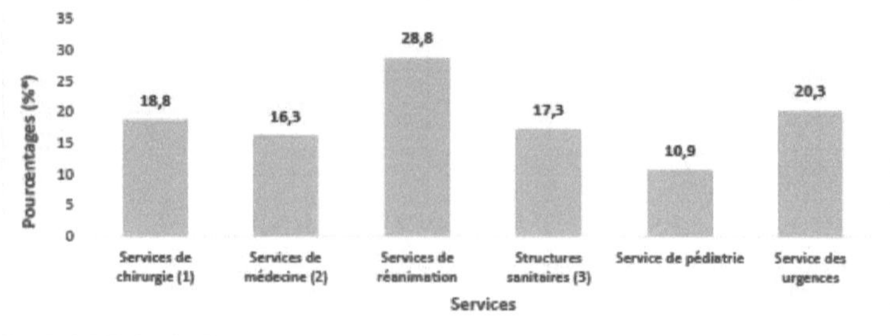

Figure 23: Distribution of MRSA in different departments

(1) Ratio of MRSA strains in the department to the total number of *S. aureus* strains tested for cefoxitin in the department concerned.

(2) Orthopaedics, Paediatric surgery, ENT, General surgery, Gynecology, Urology, Ophthalmology, Neurosurgery.

(3) Nephrology, Dermatology, Infectious diseases, Cardiology, Neonatology, Rheumatology, Pneumology, Endocrinology, Gastrology, Psychiatry, Internal medicine, Neurology, Occupational medicine.

(3): Dispensaries, regional hospitals.

Among the surgical departments, the highest rates of MRSA were found in the general surgery department (24.4%, n=76), followed by the neurosurgery department (24.3%, n=26) and the urology department (23.9%, n=17) **(Figure 24).**

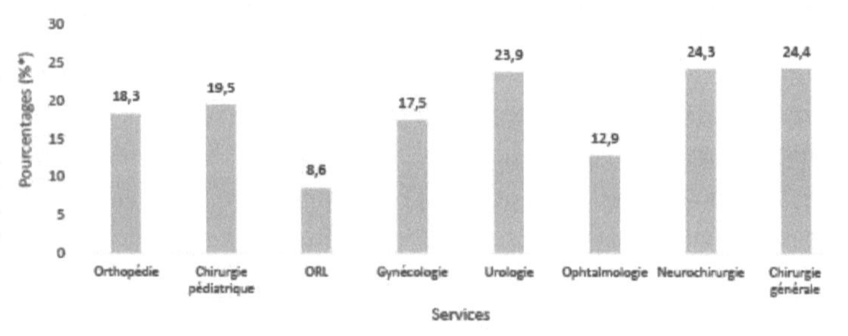

Figure 24: Distribution of MRSA in surgical departments

8.3. DISTRIBUTION OF MRSA BY TYPE OF SAMPLE

In our study, the highest percentage of MRSA was found in blood cultures (23.3%, n=186) and materials (24.5%, n=34) **(Figure 25).** This difference was statistically significant (p=0.003).

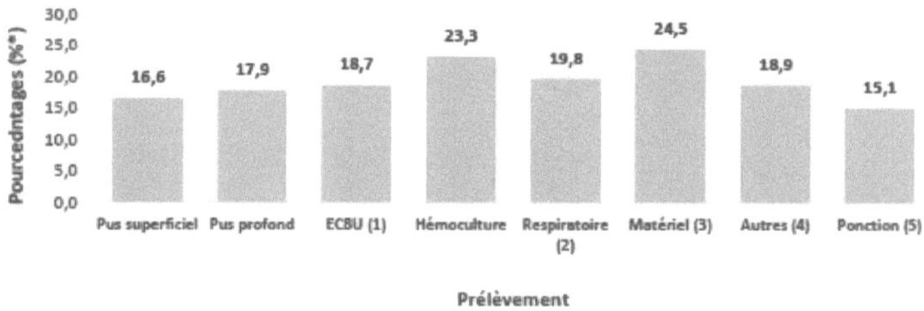

Figure 25: Distribution of MRSA by sampling site

(1) Ratio of MRSA strains in the sample to the total number of *S. aureus* strains tested for cefoxitin in the sample concerned.

(2) Cytobacteriological examination of urine.

(3) Cytobacteriological examination of sputum, Tracheal aspiration, Bronchoalveolar lavage, Protected distal swab.

(4) Catheter, Drain/Redon.

(5) Sphere ENT, Genital sampling, Ophthalmological sampling, Biopsy, Bones.

(5): Cephalo-spinal fluid puncture, ascites puncture, joint puncture.

8.4. EVOLUTION OF SARM OVER THE YEARS IN THE VARIOUS SERVICES

Over the period from 2012 to 2022, MRSA rates varied greatly between the different departments. The percentages varied between 11% (n=12) and 28.3% (n=28) in the medical departments.

In the surgical departments, these frequencies fluctuated between 12.4% (n=25) and 27.6% (n=40).

MRSA rates were stable in intensive care units, with the exception of 2014, when there was a very high peak (54%, n=27).

No MRSA was isolated from the paediatric ward in 2021. In 2020, however, this department had the highest rate of methicillin resistance (25%, n=5).

In addition, no methicillin-resistant strains were noted in the emergency department in 2013 and 2014. The highest rate was observed in 2015 (36.4%, n=4).

The health facilities had no MRSA in 2012 and the rate of MRSA varied from 5.6% (n=1) to 29.4% (n=5) **(Figure 26).**

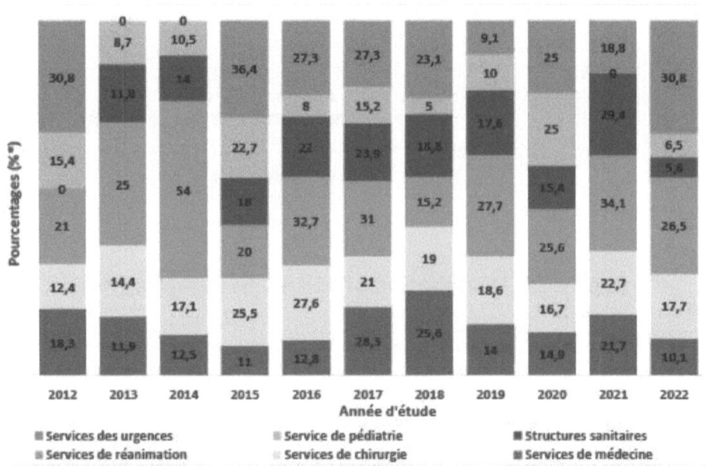

Figure 26: Changes in MRSA over the years in the various departments

*** Ratio of MRSA strains during the year of study in the department concerned to the total number of strains tested during the same year in the same department.**

8.5. COMPARISON OF RESISTANCE RATES OF MRSA AND MSSA STRAINS TO DIFFERENT ANTIBIOTICS

MRSA strains have higher rates of resistance to other families of antibiotics than SASM strains. More than half of MRSA isolates were resistant to kanamycin and cyclins. This difference was statistically significant ($p<0.05$) **(Figure 27).**

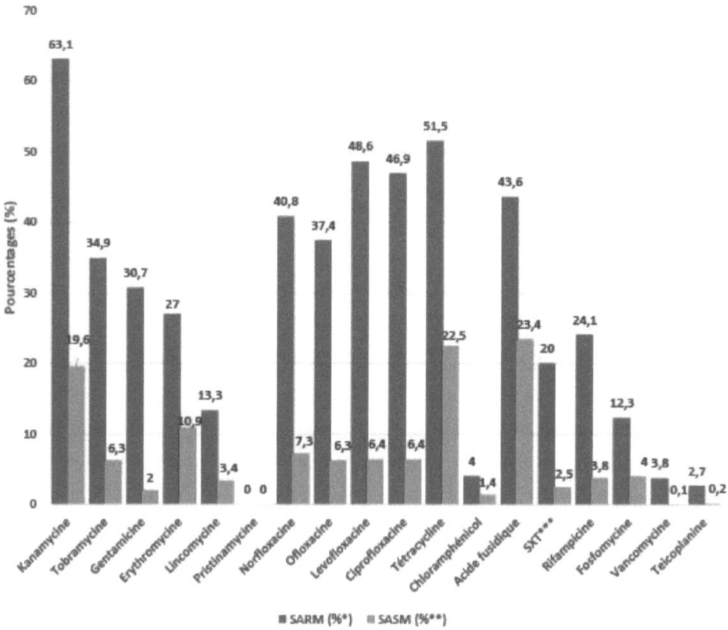

Figure 27: Comparison of resistance rates of MRSA and MSSA strains to different antibiotics

Ratio of antibiotic-resistant isolates to total number of methicillin-resistant strains

*** Ratio of antibiotic-resistant isolates to total number of methicillin-susceptible strains.**

**** Trimethoprim-sulfamethoxazole.**

4 DISCUSSION

Our study is retrospective and covers a period of 11 years (from 2012 to 2022). The objective was to study the epidemiology of *S, aureus* strains and their antibiotic resistance during this period in the microbiology laboratory of the Fattouma Bourguiba University Hospital in Monastir. The prevalence of *S. aureus* isolates was 13% (n=4541, out of a total of 34810). The highest rates were found in surgery (40.8%, n=1851), medicine (27.9%, n=1265) and intensive care (11.6%, n=527) departments. The majority of strains were isolated from pus samples (59%, n=2678), blood cultures (18.2%, n=825) and respiratory samples (9.7%, n=439). *S. aureus-positive* specimens were found mainly in surgical, medical and intensive care units, respectively. Resistance of our isolates to penicillin G and methicillin was 96.4% and 18.6%, respectively. These rates were relatively stable during the study period. The same was true for the other families of antibiotics, with the exception of glycopeptides and fusidic acid. The highest levels of resistance were found mainly in intensive care and medical wards, in material samples, punctures and ECBU. MRSA strains predominated in the intensive care unit (28.8%) and the emergency department (20.3%). There was also a predominance of MRSA in materials (24.5%) and blood cultures (23.3%). In addition, methi-R strains had higher rates of resistance to other families of antibiotics than methi-S strains.

1. DISTRIBUTION OF S. AUREUS STRAINS BY YEAR OF STUDY

In this study, the prevalence of *S, aureus* isolates was 13% (n=4541) **(Table XII).** This result is comparable with studies conducted in Nepal, Ethiopia, Kigali and Pakistan [14-17]. Other studies have reported results superior to those noted in our study [18-24]. Nevertheless, a Moroccan study published in 2015 showed a prevalence of *S, aureus* at 5% [25] **(Table XII)**. In this study, the highest prevalence of *S, aureus* was found in 2022 (19%). The lowest was in 2020 (10.7%). For the rest of the years, prevalence was relatively stable **(Table I).** The high prevalence in 2022 and the low prevalence in 2020 could be explained by the selection of these strains following the antibiotic treatment prescribed during the COVID-19 period (in 2020-2021), which was aimed more at Gram-negative bacilli. In fact, patients affected by COVID-19 were mainly superinfected by gram-negative bacilli. Prescribing antibiotics that cover these germs could therefore lead to the selection of gram-positive bacteria, including *S, aureus*. The stable prevalence of *S, aureus* was noted in the LART 2012-2015-2016-2017 data, as well as in American and European studies conducted between 2010 and 2016 [20,21,23,24,26-29]. On the other hand, according to a Moroccan study conducted between 2007 and 2014 and published in 2015, a steady increase in the number of isolates over the years was observed [25]. Also, in other studies, a constant increase in the prevalence of *S, aureus* strains, particularly between 2017 and 2019, was noted in Egypt, Ethiopia, Madagascar, Malawi, Mali, Mozambique, Nigeria, South Africa, Sudan, Tunisia, Uganda and Zambia. These data were reported to GLASS (Global Antimicrobial Resistance and Use Surveillance System), which was launched by the WHO (World Health Organisation) [30].

Table XII: Prevalence of S. aureus strains according to the literature literature

Authors	Prevalence of *S. aureus*
Sapkota et al (Nepal January 2019-June 2019) [18]	20%
Shahi et al (Nepal April 2013-September 2013) [15]	14,4%
Muhammad Mubashar Idrees (Pakistan June 2021-December 2021) [17]	10,2%
Dilnessa and Bitew (Ethiopia September 2013- April 2014) [14]	14,3%
Gahamanyi Noe (Kigali April 2014-July 2014)[16]	10%
Frikh (Morocco January 2007-December 2014) [25]	5%
Ben Lahlou (Morocco April 2012-March 2016) [19]	27,6%
ECDC (Europe 2019) [20]	20,6%
ECDC (Europe 2020) [23]	21,9%
ECDC (Europe 2021) [21]	22,1%
Blerta Kika (Albania January 2014- December 2015) [22]	34,2%
Rodrigo E. Mendes (United States 2010-2016) [24]	23,7%
This study (Tunisia 2012-2022)	13%

2. DISTRIBUTION OF *S. AUREUS* ISOLATES BY DEPARTMENT

This study showed that approximately half of the *S, aureus* strains (40.8%, n=1851) were isolated in surgical departments, the majority in the orthopaedic department (17.6%, n=799). This predominance could be explained, on the one hand, by the use of implant materials during surgery, and on the other hand, it is during the perioperative period that the risk of infection is greatest [31]. In fact, *S, aureus* is capable of forming biofilms, which are aggregations of bacterial cells adhering to medical devices, implants or other surfaces. The biofilm formation process involves the cells adhering to the host's surface polymers or extracellular matrix. The bacteria then clump together, multiply and form microcolonies. Next, the germs will synthesise an exopolysaccharide that is involved in microbial adhesion, biofilm structure and also energy reserve, giving rise to a mature biofilm. The final stage involves the detachment and dispersal of the cells, which have the capacity to adhere to new surfaces and reform other biofilms. This gives rise to a silent infection that is prolonged and becomes chronic, with the problem of eradicating these germs. Moreover, biofilm formation increases bacterial resistance to antibiotics, disinfectants and the host's immune response [32,33]. On the other hand, patients admitted to a surgical ward may become self-infected by their own endogenous commensal flora, given the frequent chronic individual carriage (estimated at 20-30%) [34]. Moreover, carriage may progress to infection either directly or via colonisation. In addition, more than % of isolates were received from medical wards (27.9%, n=1265), mainly from the nephrology ward. Given that this department includes a dialysis, nephrology and intensive care unit (NICU), the infection could be caused by the use of catheters in patients with chronic renal failure at the dialysis stage, as these seem to be a risk factor for bacteremia with a starting point of infection on equipment when they are inserted into a vein for a long time [35]. However, these results differ from other national and international studies [22,26,35-37]. Comparable results were reported in a Moroccan study conducted in 2015 [25] **(Table XIII).** A study carried out in Congo between 2019 and 2020 showed a predominance of *S. aureus* strains in neonatology [35]. In our hospital, we no longer receive samples from neonatology or onco-hematology because the laboratory of the maternity and neonatology centre became autonomous 2 years after it was founded in 2011.

Table XIII: Distribution of *S. aureus* strains according to departments according to the literature

Authors	Distribution of *S. aureus* strains by department
L. G. Gangoue et al (Congo 2019 -2020) [35].	Neonatology: 29.6% Paediatrics: 24.1% General surgery: 18.5% Infectious diseases: 14.8% Emergency: 7.4% Gyneco-obstetrics: 3.7
LART (Tunisia 2017) [26]	Medical services: 29.7% General surgery: 28.6% Paediatrics: 23.3% Intensive care: 8.6% Outpatient care: 5.6 Onco-hematology: 3.2% Gynecology: 1% Neonatology: 1
Frikh (Morocco January 2007-December 2014) [25]	Orthopaedics: 15% discount Other surgical departments: 16% Nephrology: 4% Dermatology: 4.6 Other medical services: 12.4% Intensive care: 7.7% Emergency: 3.3 Paediatrics: 2.5
Nabyi I, Tieb C. (Algerie 2018) [36]	Nephrology: 0.5% Dermatology: 14.6% Other medical departments: 31.7% Orthopaedics: 5.8% Other surgical departments: 9.7% Intensive care units: 6.8% Paediatrics: 2.4
Blerta Kika (Albania January 2014- December 2015) [22]	Surgical departments: 19.1% Nephrology: 16.9% Dermatology: 29.2% Other medical departments: 34.8
ONERBA (France 2017) [37]	Surgical departments: 11.7% Medical departments: 31.7% Onco-hematology: 4% Gynecology: 0.4 Intensive care units: 14.1% Paediatric units: 3.9 Emergency department: 24.5% Other: 9.7
This study (Tunisia 2012-2022)	Nephrology: 7.4% Dermatology: 5.3% Other medical services: 15.2 Orthopaedics: 17.6 Other surgical departments: 23.2% Intensive care anaesthesia: 8.3% Intensive care medicine: 8.3%

	Emergency: 3.3% Paediatrics: 6.4

3. DISTRIBUTION OF *S. AUREUS* ISOLATES PER SAMPLE

S, aureus is one of the main pathogens of humans, causing various suppurative diseases, food poisoning, pneumonia, septicaemia and toxic shock syndrome by producing numerous virulence factors [38]. The results of our study showed that the majority of strains were isolated from superficial pus samples (42.2%, n=1916). This compares favourably with other studies [16,18,26-29,36,39]. In contrast, a study carried out in Albania showed a predominance of *S. aureus* isolates in blood cultures, with pus samples ranking second [22]. The strains of *S, aureus* reported in our study, defined as being responsible a priori for a superficial suppurative infection, are isolated in monomircrobial culture from a superficial pus sample, which was swabbed in almost all cases. However, it is essential to compare the identification of the isolate with the clinical data, as this is a pyogenic germ that forms part of the commensal flora of the skin. So, to distinguish between commensalism and pathogenicity of the strain, you obviously need clinical information. However, in our laboratory, most samples are collected without any clinical data. A local suppurative infection with *S. aureus* is due to the expression of certain virulence factors, namely enzymes, toxins and wall constituents. In fact, this bacterium possesses surface proteins (adhesins) that enable it to adhere to components of the extracellular matrix and penetrate the body when the cutaneous-mucosal barrier is breached (cracks, abrasions, surgical incisions, burns, intravenous catheters, etc.). Once the germ has crossed the mucosa, it protects itself against the host's defences by means of its capsule, which is present in 80% of cases, protein A and coagulase, which forms the staphylothrombin complex with prothrombin. This complex leads to the polymerisation of fibrinogen into fibrin and the formation of a clot that protects the bacteria from phagocytosis. *S. aureus* also possesses factors which allow the infection to spread, namely toxins (Hemolysin a, B, 5 and Panton Valentine Leucocidin or PVL) which bind to target cells and form membrane channels through which ions can pass, coagulase and staphylokinase, which contribute to hematogenic diffusion of the bacteria from the primary focus, and other enzymes (proteases, elastase and hyaluronidase) which are capable of degrading connective tissue, thereby promoting the spread of the focus [4]. Ideally, therefore, the bacteria should be isolated at the stage of localized suppurative infection and treated, as the infection may spread and become generalized. Moreover, blood cultures ranked second in our study (18.2%, n=825) and the presence of the bacteria in the blood could be secondary to a local suppurative infection. In fact, *S. aureus* is the 2eme most frequent cause of bacteremia after Coagulase Negative Staphylococci (CNS), with a mortality rate of 20 to 40%, indicating a poor prognosis [5]. Respiratory samples ranked 3eme (9.7%, n=439). Moreover, *S. aureus* can sometimes secrete PVL, a specific leukotoxin responsible for severe necrotizing pneumonitis, with a high mortality rate of between 30 and 75% [6]. These data are comparable to the results of other studies [18,26-29,36]. However, an article from Thailand showed a different result, with isolates predominating in pus samples, respiratory samples ranking 2eme and blood cultures ranking 3eme [39]. Furthermore, according to an American study conducted between 2000 and 2014, respiratory samples ranked 1ere followed by pus samples and blood cultures [40]. In this study, respiratory samples were taken mainly from intensive care patients, most of whom were intubated and mechanically ventilated, and who received antibiotic therapy that mainly covered Gram-negative bacteria, which led to an imbalance in the flora and selection of Gram-positive bacteria. According to our results, ECBU ranked 4th (4.4%, n=199) compared with the literature [18,26,36,39]. The majority of *S. aureus-positive* ECBUs were found in health facilities and the emergency department. Given the lack of clinical data, this could be explained by possible prior antibiotic therapy, which facilitated the selection of certain germs. Although *S. aureus* accounts for only 0.5-6% of urinary tract infections, untreated infection can lead to serious, life-threatening conditions. Isolation of S. aureus from urine samples should be investigated further to exclude bacteremia (e.g. endocarditis) [41] **(Table XIV).**

Table XIV: Distribution of S. aureus strains per sample according to the literature

Authors	Distribution of *S. aureus* strains by type of sample
Sapkota et al (Nepal January 2019-June 2019) [18]	Superficial pus: 56.4% Deep pus: 23.3% ECBU: 2.3% Blood culture: 11.3% Respiratory: 4.5% Material: 2.3
LART (Tunisia 2017) [26]	Pus: 47.6%. ECBU: 6.2% Blood culture: 15% Respiratory: 14.7% Puncture: 5.5
LART (Tunisia 2012) [27]	Pus : 48.9% ECBU: 5.9% Blood culture: 17.7% Respiratory: 11.2% Puncture: 5.1
LART (Tunisia 2015) [28]	Pus : 43.9% ECBU: 5.7% Blood culture: 21.4% Respiratory: 13.8% Puncture: 5.1
LART (Tunisia 2016) [29]	Pus : 46.9% ECBU: 8% Blood culture: 18.9% Respiratory: 14% Puncture: 4.2

P. WAITAYANGKOON ET AL. (Thailand January 2017- December 2017)[39]	Pus: 44%. ECBU: 5.2% Blood culture: 10% Respiratory: 30%
Nabyi I, Tieb C. (Algerie 2018) [36]	Pus: 58% ECBU: 3.5% Blood culture: 31.7% Material: 4.2% Puncture: 2.6
Gahamanyi Noe (Kigali April 2014-July 2014)[16]	Pus: 63.3% Blood culture: 36.7
Blerta Kika (Albania January 2014- December 2015) [22]	Pus : 26.2% ECBU: 17.2% Blood culture: 29.5% Respiratory: 14.8
Sanjat Kanjila (United States 2000-2014) [40]	Pus: 28 Blood cultures: 16 Respiratory: 34 Other: 22
This study (Tunisia 2012-2022)	Superficial pus: 42.2% Deep pus: 16.8% ECBU: 4.4% Blood culture: 18.2% Respiratory: 9.7% Material: 3.2% Puncture: 3.7

4. STUDY OF THE RESISTANCE OF STRAINS *of 5. AUREUS* TO ANTIBIOTICS

S, aureus is naturally resistant to certain antibiotics, namely temocillin, mecillinam, aztreonam, nalidixic acid and colistin [42,43]. In addition to natural resistance, this bacterium is capable of developing different mechanisms to neutralise the action of antibacterial agents. Multi-antibiotic resistance in staphylococci is a growing problem in clinical practice and the choice of antimicrobial agents has become increasingly limited as this germ continues to develop resistance to a range of antimicrobial agents [44,45].

4.1. RESISTANCE OF *S. AUREUS* ISOLATES TO PENICILLIN G

Strains resistant to penicillin G appeared in the 1950s, after penicillin G was introduced to the market around 1940. They were able to hydrolyse the antibiotic by synthesising an enzyme called **penicillinase**, which was capable of breaking the в lactam cycle [46]. These bacteria therefore acquire resistance to penicillin G, penicillin V, aminopenicillins, carboxypenicillins and ureidopenicillins [42]. To overcome this resistance mechanism, pharmaceutical laboratories synthesised new penicillins in the 1960s. These are the semi-synthetic penicillins in group M [46,47]. In this study, the results obtained showed a very high level of resistance in *S. aureus* strains to penicillin G (96.4%, n=4296) compared with the rates reported in other studies conducted both nationally and internationally [10,13,26,37,44, 48-50] **(Table XV).** This could be explained by the ease of access to this antibiotic and its use without a prior medical prescription. In addition, clinicians use these molecules as empirical therapy, which will lead to selection pressure on this bacterial population, resulting in the appearance of resistant strains [13].

Table XV: Resistance rates of *V. aureus* strains to penicillin G according to the literature

Authors	Resistance rate
L. Chelkeba and T. Melaku (Ethiopia 1988-2020) [13]	84%
F. R. Koinam et al (BURKINA FASO 2016) [10]	92%
Alfeky et al (Egypt September 2017-December 2018)[48]	81,2%
Naimi et al (Afghanistan September 2016-February 2017) [44]	99%
LART (Tunisia 2017) [26]	94,7%
ONERBA (France 2017) [37]	85,9%
Coombs GW (Australia 2020) [49].	76,3%
McNeil JC et al (United States 2019) [50].	79,6%
This study (Tunisia 2012-2022)	96,4%

4.2. RESISTANCE OF S. AUREUS ISOLATES TO METHICILLIN

Methicillin-resistant *S. aureus* is a major cause of community and hospital infections worldwide. The first epidemic was described in the 1960s. This resistance to methicillin is essentially due to the expression of an additional penicillin-binding protein (PLP2a) encoded by the mecA gene, which is characterised by its low affinity for penicillin. However, there are other genes involved in the expression of this resistance. The Centers for Disease Control and Prevention (CDC) have classified MRSA as a "serious threat" [11,48,51].

In this study, MRSA strains accounted for 18.6% (n=828). Similar rates have been reported in LART 2017 and AGAR (Australian Group on Antimicrobial Resistance) data [26,49]. In addition, other studies have demonstrated different frequencies of MRSA isolation to that reported in this work [24,25,40,43,44,48,52-58]. Given that MRSA is transmitted mainly by direct contact with the contaminated hands of a carrier or infected person or those of healthcare staff, or with contaminated surfaces and objects, this unequal distribution of MRSA rates between the different studies could be explained by the application of isolation and hygiene measures for patients with MRSA, which differ between the different healthcare structures. In fact, the application of standard precautions during a care procedure or during a patient examination is compulsory, but certain health facilities may not apply them correctly, probably due to a lack of resources. In addition, given that breaking the skin barrier favours MRSA infection, this could be due to the increasing impact of invasive procedures and the overuse of broad-spectrum antibiotics [53]. Prolonged hospital stays and the prescription of broad-spectrum antibiotics are among the risk factors for MRSA [59]. The difference in the procedures to be followed in each hospital could explain this variation in the frequency of isolation of methi-R strains. Furthermore, resistance to methicillin was stable during the period of our study, compared with other studies [20,23,25-27] **(Table XVI).**

Table XVI: Comparative table of resistance rates of *S. aureus* strains to methicillin according to literature data

Authors	Resistance rate
Qodrati et al (Iran September 2018-September 2019) [53]	37,5%
Naimi et al (Afghanistan September 2016- February 2017) [44]	56,2%
Raut et al (Nepal September 2015- February 2016) [52].	43,6%
A. Khanal, S. G.C., A. Gaire et al (Nepal January 2008- August 2020) [54]	38,2%
LART (Tunisia 2017) [26]	19,3%
Frikh (Morocco January 2007-December 2014) [25]	8,7%
Alfeky et al (Egypt September 2017- December 2018) [48]	81,2%
Sara Romano-Bertrand (France 2021) [43]	12,3%
Daniela Talapan (Romania January 2017- December 2022) [55]	39,1%
Adriano La Vecchia (Italy 2017-2021) [57]	30%
MARTA KLOS (Poland 2009-2013) [58]	6%
Rodrigo E. (United States 2010-2016) [24]	10,4%
Sanjat Kanjila (United States 2000-2014) [40]	49,9%
Margaret Carre (United States 2010-2019) [56]	45,3%
Coombs GW (Australia 2020) [49].	17,6%
This study (Tunisia 2012-2022)	18,6%

4.3. RESISTANCE OF S. AUREUS ISOLATES TO AMINOGLYCOSIDES

Aminoglycosides are broad-spectrum antibiotics produced by strains of Streptomyces. This explains why certain molecules are reserved for use in hospitals to combat severe bacterial infections. These antimicrobial agents act on messenger ribonucleic acid (mRNA) translation and protein synthesis, specifically on the 30S subunit of the bacterial ribosome. The main mechanism of resistance to aminoglycosides is structural modification of these antibiotics by Aminoglycoside Modifying Enzymes (AMEs) [36,60]. In our series, resistance rates to kanamycin, tobramycin and gentamicin were 28%, 11.5% and 7.4%, respectively. National and international studies have shown resistance rates to kanamycin comparable to those reported here [26,29,49,55,58]. Lower levels of resistance have been observed in other studies [61,62]. According to LART 2015, the percentage of resistance to kanamycin was 75%. As for tobramycin and gentamicin, higher rates have been reported in the literature [14,26,28,29,44,48,61,63]. According to 2018 ONERBA data, gentamicin and tobramycin had resistance rates of 1.7% and 5.8% respectively [62]. However, according to a study conducted in Burkina Faso, no gentamicin resistance was observed and 16% of strains were resistant to tobramycin [10]. Aminoglycoside resistance was stable over the eleven years of the study, with the exception of kanamycin, which showed a higher figure in 2020 (42.9%, n= 39/91) **(Table XVII).** This could be explained by the lower number of strains tested compared with the other years.

Table XVII: Rate of resistance of *S. aureus* strains to aminoglycosides according to the literature

Authors	Taui GM	t of resista TB	nce И9
Naimi et al (Afghanistan September 2016- February 2017) [44]	25,7%	32,4%	
Dilnessa and Bitew (Ethiopia September 2013- April 2014) [14]	13,4%		
Ms ALLE AKAKPO Amavi Essenam (Mali 2000-2020) [61]	17 %		15 %
F. R. Koinam et al (BURKINA FASO 2016) [10]	0 %	16%	
A.-C. Dikoumba et al (Gabon 2016-2018) [63]	18 %	15%	
Alfeky et al (Egypt September 2017- December 2018) [48]	57,6 %		
LART (Tunisia 2015) [28]	34,7%	46%	75%
LART (Tunisia 2016) [29]	15%	21,7%	32,5%
LART (Tunisia 2017) [26]	12,3%	15,2%	28,7%
ONERBA (France 2018) [62]	1,7%	5,8%	6,6%
MARTA KLOS (Poland 2009-2013) [58]	6%	16,8%	
Daniela Talapan (Romania January 2017- December 2022) [55]	5,9%		
Coombs GW (Australia 2020) [49].	2%		
This study (Tunisia 2012-2022)	7,4 %	11,5%	28%

GM = Gentamicin; TB = Tobramycin; KN = Kanamycin.

4.4. RESISTANCE OF *S. AUREUS* ISOLATES TO MACROLIDES

Macrolides, lincosamides and streptogramins (MLS) are a chemically distinct group of antibiotics that act primarily by inhibiting bacterial protein synthesis. These antimicrobial agents are used as alternative medicines for the treatment of certain *5. aureus* infections, particularly in people allergic to penicillin. Macrolides and lincosamides have bacteriostatic activity against Staphylococci, while streptogramins (pristinamycin, quinupristin- dalfopristin) are composed of 2 A and B subunits, acting synergistically and possessing bactericidal activity. However, the widespread use of these antibiotics has encouraged the development of resistant strains. Three main mechanisms are involved in the bacterial resistance of *5. aureus* to macrolides, namely target modification, enzymatic inactivation and active efflux [64]. In this study, resistance rates of 13.8% and 5.3% to erythromycin and lincomycin, respectively, were demonstrated. These results are consistent with data from LART 2017 and AGAR [26,49]. However, other national [28,29] and international studies [10,14,37,44,55,58,61,62,65] have noted significantly higher rates of resistance to these two antimicrobial agents. On the other hand, no resistance to the combination of streptogramin A and B has been reported. This result is confirmed by other studies [26,28,29,66]. However, according to a meta-analysis conducted in Mali between 2000 and 2020 and data from ONERBA 2017, resistance rates to pristinamycin are 27% and 1.5% respectively [37]. In fact, there are three streptogramin resistance phenotypes:

- The SA phenotype involves an efflux system for the A component of synergistins and is encoded by the vga and vgaB genes located on plasmids.
- The SA+SB phenotype is the result of an accumulation of different mechanisms such as the vga, vgt and vgaB genes in association with methylase genes.
- The LSA phenotype, the mechanism of which and the genes encoding it are still unclear [67].

The appearance of pristinamycin-resistant strains could be explained by the selection pressure generated by the abusive and often incorrect use of this antibiotic. Failure to observe hygiene rules could therefore contribute to the human-to-human transmission of resistant strains. In Tunisia, the first pristinamycin-resistant strains of *S, aureus* were described in the Sfax microbiology department in 2014 [68-70] **(Table XVIII)**.

Table XVIII: Resistance rates of *S. aureus* strains to macrolides according to the literature

Authors	Resistance rate		
	Eryth	Linco	Prist/QDF
Shariati et al (Iran 2018) [66]			0,1 %
Naimi et al (Afghanistan September 2016- February 2017) [44]	63,8 %		
M.Khodabandeh et al (Iran 2016-2017) [65]	91,5 %	76,3 %	
Dilnessa and Bitew (Ethiopia September 2013- April 2014) [14]	53,1 %	11,9 %	
F. R. Koinam et al (BURKINA FASO 2016) [10]	44 %		
Ms ALLE AKAKPO Amavi Essenam (Mali 2000-2020) [61]	44 %	35 %	27 %
LART (Tunisia 2015) [28]	37,7 %	14,5 %	0 %
LART (Tunisia 2016) [29]	25 %	9,4 %	0 %
LART (Tunisia 2017) [26]	17,4 %	6,2 %	0,4 %
ONERBA (France 2018) [62]	27,8 %		

ONERBA (France 2017) [37]	27,6%	7,9%	1,5%
Daniela Talapan (Romania January 2017- December 2022) [55]	50%		
Rodrigo E. (United States 2010-2016) [24]	56,2%	17,7%	
Coombs GW (Australia 2020) [49].	12,6%	12,5%	
This study (Tunisia 2012-2022)	13,8 %	5,3 %	0 %

Eryth= erythromycin; Linco =Lincomycin; Prist = pristinamycin; QDF = Quinupristin and dalfopristin.

4.5. RESISTANCE OF *S. AUREUS* ISOLATES TO FLUOROQUINOLONES

S. aureus is naturally sensitive to fluoroquinolones, which inhibit bacterial growth by stopping DNA replication. Acquired resistance to this family of antibiotics is crossover between the different molecules, known as a "class phenomenon", and is secondary to the accumulation of several "staircase" resistance mechanisms. Most bacterial resistance is chromosomal, due to chromosomal point mutations in the target (gyrase and/or topoisomerase IV) or, more rarely, to an efflux system. In addition, this resistance could be plasmidic (the Qnr plasmid) [42,71,72]. In this study, resistance to ciprofloxacin, ofloxacin, levofloxacin and norfloxacin was 14.5%, 12.2%, 14.6% and 13.3%, respectively. These percentages may be biased, as the number of strains tested differs from one fluoroquinolone to another. Moreover, according to EUCAST recommendations, norfloxacin is used as a screening test for resistance to this therapeutic class. Indeed, if the strain is sensitive to norfloxacin, it would be sensitive to all fluoroquinolones. However, resistance to norfloxacin requires testing for sensitivity to the fluoroquinolone prescribed. Higher rates of resistance have been reported in other studies [10,14,15,24,35,44,48,61,71]. Data from LART 2012, 2015, 2016 and 2017 as well as a study carried out in Gabon and published in 2021 showed resistance rates to ofloxacin comparable to the rate reported in this study [26-29,63] **(Table XIX).** These results could be explained by the impact of previous prescription of these antimicrobial agents as monotherapy, which favours the selection of resistant mutants and leads to therapeutic failure in the event of subsequent treatment with these antibiotics. Hence the need for well-managed and justified prescribing of fluoroquinolones [72,73].

Table XIX: Resistance rates of *S. aureus* strains to fluoroquinolones according to literature data

Authors	Resistance rate			
	Cip	Oflo	Lev	Nor
Naimi et al (Afghanistan September 2016- February 2017) [44]	55,3%			
L. G. Gangoue et al (Congo 2019 -2020) [35].	67,5%	72,5%	62,5%	70%
Ms ALLE AKAKPO Amavi Essenam (Mali 2000-2020) [61]	29 %	25%	50%	

F. R. Koinam et al (BURKINA FASO 2016) [10]	40 %			
A.-C. Dikoumba et al (Gabon January 2016- March 2018) [63]		10 %		
Dilnessa and Bitew (Ethiopia September 2013- April 2014) [14]	31,4%			
Shahi et al (Nepal April 2013-September 2013) [15]	32,1%		22,9%	
Gade and Qazi (India November 2009- December 2011) [71]	57,6%	51,2 %	30,4%	
Alfeky et al (Egypt September 2017- December 2018) [48]	31,8%		37,1%	
LART (Tunisia 2017) [26]		10,2%		
LART (Tunisia 2012) [27]		13,5%		
LART (Tunisia 2015) [28]		13,8%		
LART (Tunisia 2016) [29]		10%		
Rodrigo E. (United States 2010-2016) [24]			38,7%	
Rodrigo E. (Europe 2010-2016) [24]			26,9%	
This study (Tunisia 2012-2022)	14,5%	12,2%	14,6%	13,3%

Cip = Ciprofloxacin; Oflo = Ofloxacin; Lev = Levofloxacin; Nor = Norfloxacin.

4.6. RESISTANCE OF S. AUREUS ISOLATES TO GLYCOPEPTIDES

Glycopeptides are essential weapons in the fight against MRSA, particularly vancomycin, which has remained a reference treatment for more than 50 years since its introduction. In some cases, teicoplanin may become an alternative to vancomycin due to the low incidence of adverse effects such as nephrotoxicosis, its lower cost and because it remains active against certain vancomycin-resistant strains, such as vanB-type strains [34,74,75]. According to the national recommendations issued in February 2006, teicoplanin is the treatment of choice for osteoarticular infections, as it penetrates bone more easily and is easier to administer as a single daily dose, either as a short infusion or intramuscularly [34]. The first vancomycin-intermediate *S. aureus* and the first vancomycin-resistant *S. aureus* were detected, respectively, in Japan in 1997 and in the United States in 2002. However, the exact mechanism of resistance remains poorly elucidated [67, 74]. The *S. aureus* strains included in this study show excellent sensitivity to glycopeptides, with only 1.2% and 0.9% of strains resistant to vancomycin and teicoplanin, respectively. Some other studies show that the sensitivity of *S, aureus* to glycopeptides is low or even unalterable [24,26-29,37,44,49,55,63). Higher rates of resistance have been reported in other articles [10,13,14,48]. According to the Centers for Disease Control and Prevention (CDC), the risk factors for the emergence of VRSA are: previous infection or colonisation with MRSA and enterococcus, underlying medical conditions (such as chronic skin ulcers and diabetes) and previous treatment with vancomycin [76]. This could explain the significant increase in resistance to vancomycin in 2021 and to teicoplanin in 2020 and 2021 (**Table XX).**

Table XX: Rate of resistance of *S. aureus* strains to glycopeptides according to the literature

Authors	Rate of resistance Van	Teico
Legese Chelkeba (Ethiopia 1988- March 2020) [13]	29%	
Naimi et al (Afghanistan September 2016- February 2017) [44]	0%	
Dilnessa and Bitew (Ethiopia September 2013- April 2014) [14]	5,1 %	
F. R. Koinam et al (BURKINA FASO 2016) [10]	4 %	
A.-C. Dikoumba et al (Gabon January 2016- March 2018) [63]	0 %	0 %
Alfeky et al (Egypt September 2017- December 2018) [48]		17,1 %
LART (Tunisia 2017) [26]	0 %	0 %
LART (Tunisia 2012) [27]	0 %	0 %
LART (Tunisia 2015) [28]	0,2 %	0,2 %
LART (Tunisia 2016) [29]	0,5 %	0 %
Coombs GW (Australia 2020) [49].	0%	0,1%
Rodrigo E. (United States 2010-2016) [24]	0%	0,2%
Rodrigo E. (Europe 2010-2016) [24]	0%	0,1%
ONERBA (France 2017) [37]	0,3%	0,3%
Daniela Talapan (Romania January 2017- December 2022) [55]	0%	0,2%
This study (Tunisia 2012-2022)	1,2 %	0,9 %

Van: vancomycin; Teico: Teicoplanin.

4.7. Resistance of *S. aureus* isolates to other antibiotics

4.7.1. Fosfomycin

Fosfomycin is a bactericidal antimicrobial agent with a broad spectrum of activity against Gram-negative and Gram-positive pathogens. The principal mechanism of resistance is alteration of the antibiotic's bacterial transport systems as a result of chromosomal mutation. Because of its availability for oral treatment and its lower cost than other anti-staphylococcal agents, fosfomycin has great prospects for application in the treatment of MRSA [77,78]. In this study, resistance to fosfomycin was 5.6%. Comparable rates have been reported in the literature [27,36]. Furthermore, in two studies conducted in Gabon and Ghana, no resistance to this antimicrobial agent was detected [63,79]. However, a 23% resistance rate was reported in a Malian study also published in 2021 [61] (**Table XXI**). Given that the use of fosfomycin as a monotherapy very rapidly selects for resistance, these results could be explained by its inappropriate and often ill-considered prescription [73].

Table XXI: Resistance rates of *5. aureus* strains to fosfomycin according to the literature

Authors	Resistance rate
Ms ALLE AKAKPO Amavi Essenam (Mali 2000-2020) [61]	23%

A.-C. Dikoumba et al (Gabon January 2016- March 2018) [63]	0%
Nana Ama Amissah (Ghana 2014-2015) [79]	0%
LART (Tunisia 2012) [27]	5,2%
Nabyi I, Tieb C. (Algerie 2018) [36]	7,37
This study (Tunisia 2012-2022)	5,6%

4.7.2. FUSIDIC ACID

Fusidic acid is derived from the fungus Fusidium coccineum. This antimicrobial agent has moderate activity against most Gram-positive bacteria, including staphylococci, MRSA and certain Gram-negative anaerobic bacteria. It is a very important and valuable alternative to vancomycin in the fight against resistant organisms. Fusidic acid has mainly bacteriostatic effects, although it can be bactericidal in high concentrations. It is characterised by a high mutation rate [68,80]. In this study, we noted an overall rate of resistance to this antibiotic of 27.1%, with a constant increase in this rate over the years, reaching 40.6% (n=191) in 2022. These results could be explained by overuse of the drug, failure to adhere to treatment times and prescription as a monotherapy, which leads to the selection of resistant mutants. Moreover, the French National Authority for Health (HAS) has indicated that local treatment with fusidic acid promotes resistance, which could be detrimental to its use by the oral or injectable route [73,81]. Lower rates have been reported in the literature [10,37,49,62,79]. However, an Algerian study conducted in 2019 showed a higher resistance rate of 43.83% [36] **(Table XXII)**.

Table XXII: Resistance rate of *S. aureus* strains to fusidic acid fusidic acid according to literature data

Authors	Resistance rate
F. R. Koinam et al (BURKINA FASO 2016) [10]	4 %
Nana Ama Amissah (Ghana 2014-2015) [79]	0%
Nabyi I, Tieb C. (Algerie 2018) [36]	43,83 %
ONERBA (France 2018) [62]	17,8 %
ONERBA (France 2017) [37]	5,4%
Coombs GW (Australia 2020) [49].	3%
This study (Tunisia 2012-2022)	27 ,1 %

4.7.3. TETRACYCLINE

These are inexpensive, broad-spectrum antibiotics widely used to treat *S. aureus* infections in humans [82]. The results of our study showed a rate of resistance to this antimicrobial agent of 28.9%, compared with rates reported in other national and international studies [26-29,44,58]. Higher rates have been reported in other studies [10,13,48,61,63]. However, according to a study carried out in the United States between 2010 and 2016 and published in 2019, the rates of tetracycline resistance in America and Europe were 6.6% and 8%, respectively.

A percentage of 2.95% was noted in an Algerian study conducted in 2018 [36] **(Table XXIII)**.

Table XXIII: Resistance rates of *5. aureus* strains to tetracycline according to the literature

Authors	Resistance rate
Legese Chelkeba (Ethiopia 1988- March 2020) [13]	61%
Naimi et al (Afghanistan September 2016- February 2017) [44]	22,8%
Ms ALLE AKAKPO Amavi Essenam (Mali 2000-2020) [61]	68
F. R. Koinam et al (BURKINA FASO 2016) [10]	48%
A.-C. Dikoumba et al (Gabon January 2016- March 2018) [63]	33%
Alfeky et al (Egypt September 2017- December 2018) [48]	52,1%
LART (Tunisia 2017) [26]	32%
LART (Tunisia 2012) [27]	30,6%
LART (Tunisia 2015) [28]	29,1%
LART (Tunisia 2016) [29]	29,2%
Nabyi I, Tieb C. (Algerie 2018) [36]	2,9%
MARTA KLOS (Poland 2009-2013) [58]	16,8%
Rodrigo E. (Europe 2010-2016) [24]	8%
Rodrigo E. (United States 2010-2016) [24]	6,6%
This study (Tunisia 2012-2022)	28,9%

4.7.4. TRIMETHOPRIM-SULFAMETHOXAZOLE COMBINATION

Trimethoprim is used in combination with sulphonamides. It potentiates their action by inhibiting folinic acid synthesis. This combination has a bactericidal effect. It is a possible alternative for the treatment of MRSA [83]. The timethoprim-sulfamethoxazole combination retains very good activity against all the strains tested, with 5.9% of resistant strains. This rate is comparable to data from LART 2012, 2015, 2016 and 2017 and to the results of a study conducted in Gabon in 2021 [26-29,63]. However, other international studies have reported higher rates [10,36,44,48,61]. An American study published in 2018 and data from ONERBA 2017 and 2018 showed lower resistances of 1.7%, 1% and 0.4%, respectively [62]. **(Table XXIV)**

Table XXIV: Rate of resistance of strains *of 5. aureus* to the combination timethoprim-sulfamethoxazole combination according to literature data

Authors	Resistance rate
Naimi et al (Afghanistan September 2016- February 2017) [44]	42,9%
Ms ALLE AKAKPO Amavi Essenam (Mali 2000-2020) [61]	41%
F. R. Koinam et al (BURKINA FASO 2016) [10]	60
A.-C. Dikoumba et al (Gabon January 2016- March 2018) [63]	6%
Alfeky et al (Egypt September 2017- December 2018) [48]	27,1%
Nabyi I, Tieb C. (Algerie 2018) [36]	25%
LART (Tunisia 2017) [26]	2,2%
LART (Tunisia 2012) [27]	2,5%

LART (Tunisia 2015) [28]	3,4%
LART (Tunisia 2016) [29]	3,1%
Rodrigo E. (United States 2010-2016) [24]	1,7%
ONERBA (France 2018) [62]	0,4%
ONERBA (France 2017) [37]	1%
This study (Tunisia 2012-2022)	5,9%

4.7.5. RIFAMPICIN

It is the broad-spectrum antibiotic most widely used in clinical practice. It is a powerful bactericidal agent, characterised by very good tissue diffusion. This antimicrobial agent is used in combined therapy for both material-borne infections and serious *S. aureus* infections. Thus, this antibiotic can improve the therapeutic response to infection by *S. aureus* with Intermediate Vancomycin Susceptibility (VISA) when used in combination with linezolid or daptomycin [84,85]. In our series, 7.4% were resistant to rifampicin. This result is consistent with some data in the literature [10,36,44]. Other international studies have shown lower rates of resistance [20,49,63]. On the other hand, an Egyptian study conducted in 2022 and data from ONERBA-France 2018 reported resistance rates of 24.8% and 18.1% respectively [48,62]. The long-term use of rifampicin and its prescription as a monotherapy could explain the emergence of strains resistant to this antibiotic [85] (**Table XXV**).

Table XXV: Resistance rates of *S. aureus* strains to rifampicin according to the literature

Authors	Resistance rate
Naimi et al (Afghanistan September 2016- February 2017) [44]	4,8%
Alfeky et al (Egypt September 2017- December 2018) [48]	24,8%
Nabyi I, Tieb C. (Algerie 2018) [36]	6,63%
F. R. Koinam et al (BURKINA FASO 2016) [10]	4%
A.-C. Dikoumba et al (Gabon January 2016- March 2018) [63]	2%
ECDC (Europe 2019) [20]	0,5 %
ONERBA (France 2018) [62]	18,1%
Coombs GW (Australia 2020) [49].	0,3%
This study (Tunisia 2012-2022)	7,4%

4.7.6. CHLORAMPHENICOL

It is a broad-spectrum bacteriostatic agent that interferes with protein synthesis by binding to the 50S ribosomal subunit. There are three mechanisms of resistance to this antibiotic, namely enzymatic inactivation, active efflux or target modification [86]. The percentage of resistance to chloramphenicol in this work was 1.9%, compared with data from LART 2012, 2015, 2016 and 2017 [26-29]. However, this rate remains lower compared with the results of some international studies [10,13,14,44,48,61]. This low rate of resistance could be explained by the limited use of this antibiotic therapy. The widespread use of chloramphenicol has been hampered by its toxicity, as systemic administration is associated with irreversible aplastic anaemia. Its use is currently limited mainly to topical applications in cases of staphylococcal conjunctivitis [86] **(Table XXVI).**

Table XXVI: Resistance rates of *S. aureus* strains to chloramphenicol according to literature data

Authors	Resistance rate
Legese Chelkeba (Ethiopia 1988- March 2020) [13]	49%
Naimi et al (Afghanistan September 2016- February 2017) [44]	35,3%
Dilnessa and Bitew (Ethiopia September 2013- April 2014) [14]	18,6%
Ms ALLE AKAKPO Amavi Essenam (Mali 2000-2020) [61]	50%
F. R. Koinam et al (BURKINA FASO 2016) [10]	8%
Alfeky et al (Egypt September 2017- December 2018) [48]	25,3%
LART (Tunisia 2017) [26]	32%
LART (Tunisia 2012) [27]	30,6%
LART (Tunisia 2015) [28]	29,1%
LART (Tunisia 2016) [29]	29,2%
This study (Tunisia 2012-2022)	1,9%

5. MRSA EPIDEMIOLOGY

5.1. DISTRIBUTION OF MRSA TO DIFFERENT DEPARTMENTS

MRSA infections continue to increase worldwide in hospitals and, more recently, in the community. In fact, transmission of these micro-organisms is very often carried from essentially human reservoirs, consisting of infected or colonised patients and hospital staff, and even families [87]. In this study, the highest rate of MRSA was found in the intensive care unit (28.8%), followed by the emergency department (20.3%), surgical departments (18.8%), mainly in the general surgery department, and health facilities (17.3%). The percentage of MRSA in other specialities falls within this range, around an average of 13.6%. The high rate of MRSA in the intensive care unit and emergency department could be explained by its ability to be present in the biofilm of devices such as catheters, intubation tubes and urinary catheters. The biofilm favours the transition to chronicity and facilitates the persistence and multiplication of MRSA on the site and therefore the acquisition of resistance and the transfer of this resistance between organisms, which represents an obstacle to the action of antibiotics. In surgical departments, MRSA infection can be acquired preoperatively, intraoperatively or postoperatively. Contamination could occur from the patient himself via the hematogenous or cutaneous route, in the event of prior colonisation by *S', aureus* or in the event of active infection concomitant with the operation. The operator could also be implicated, through his hands, skin flakes, aerocontamination or the use of septic instruments. Post-operative infection could be secondary to contamination by soiling of the scar or by the hematogenic route, from equipment such as a catheter, drain or urinary catheter [88]. MRSA in healthcare facilities is part of a community infection and is most often secondary to contact with contaminated surfaces [89]. The studies carried out on the epidemiology of MRSA are numerous and varied. Some of them have been carried out on a national scale, while others have been conducted on an international scale. This makes them difficult to compare. A Moroccan study published in 2015 showed that MRSA strains were isolated mainly from burns units (20.4%), trauma units (16.3%) and intensive care units (10.4%) [25]. Another Algerian study showed different results. MRSA strains were isolated mainly in dermatology (14.6%), the burns centre (11.7%) and the infectious diseases department (9.7%) [36]. In addition, data from LART 2012, 2015, 2016 and 2017 showed a predominance of MRSA in the intensive care unit (25%, 44.1%), the neonatology unit (38.1%, 41.9%), the paediatrics unit (17.8%, 20.6%) and the surgery unit (15.9, 19.4%) [26-29]. The difference in antibiotic resistance from one series to another within the different hospital departments could be explained mainly by the microbial ecology of the departments concerned and the implementation of a targeted, personalised policy to combat the spread of MRSA, specific to each department. To the same extent, improvements linked to the implementation of a carrier screening strategy and the promotion of prudent antibiotic therapy management have made it possible to reduce the average proportion of MRSA [20,88,90].

5.2. DISTRIBUTION OF MRSA IN DIFFERENT SAMPLES

In this study, MRSA strains were predominant in materials (24.5%, n=34) and blood cultures (23.3%, n=186). People in good health carry the bacteria on the skin or in the respiratory tract, with a preference for the nose. Reported rates of nasal carriage are around 25-30%. This does not constitute a direct risk of infection, but this changes in the event of surgery or the insertion of catheters, urinary catheters or other medical devices that are not properly maintained [39,91]. The pathogenesis of infections on materials is closely linked to the ability of *S, aureus* to produce a biofilm. This structure is a heterogeneous organisation of bacteria surrounded by an extracellular matrix. Biofilm formation takes place in several stages. The germs must first adhere to a surface. Then they clump together, multiply and form microcolonies. Next, they produce an extracellular matrix to form a mature biofilm. The bacteria can then detach from the biofilm to colonise other surfaces at a distance [33,92]. On the one hand, the biofilm enables pathogens to evade the host's defences through its mechanical barrier effect or by degrading neighbouring neutrophil polynuclears and neutralising their bactericidal effect. On the other hand, it constitutes a therapeutic obstacle due to its structure, which limits the diffusion of antibiotics, the high bacterial density within it, which reduces the efficacy of certain therapeutic classes, and the formation of *S. aureus* variant strains with small colonies and greater tolerance to antibacterial agents. This explains why few antibiotics remain active on biofilms, namely rifampicin, which is favoured for the treatment of osteoarticular infections, particularly on material [92,93]. An infection on material can be the gateway to bacteremia. In fact, *S. aureus* has several virulence factors that allow the infection to spread, namely toxins (Hemolysin a, в, 5 and Panton Valentine Leucocidin or PVL), coagulase, staphylokinase and other enzymes (proteases, elastase and hyaluronidase) [4]. Moreover, the use of a central venous catheter for a prolonged period in excess of 30 days significantly increases the risk of MRSA bacteremia [7,94]. Nonetheless, pus samples were the majority in some other studies [11,14,57,95]. In addition, a Thai study published in 2020 showed a predominance of MRSA in respiratory samples [39]. In contrast, an Egyptian study conducted between September 2017 and December 2018 showed that MRSA strains predominated in blood cultures [48]. ECBU ranked first in an Iranian study published in 2022 [53]. The presence of MRSA in blood cultures could be secondary to the use of catheters, urinary catheters or other medical devices that are not properly maintained and if hygiene procedures are not rigorously followed [81] **(Table XXVII)**.

Table XXVII: Distribution of MRSA in different samples

Authors	Resistance rate					
	Pus sup.	More prof.	HC	Res.	ECBU	Mat.
La Vecchia et al (Italy January 2017-December 2021) [57].	57		17	16	5	
Raut et al (Nepal September 2015- February 2016) [52].	33,3	42,9	49,2			
Qodrati et al (Iran September 2018-September 2019) [53]	39,1		39,9	41,7	54,2	38,5
S. AL MUSAWI (Saudi Arabia 2009- 2019) [11]	52,4		6,9	1,7		
Alfeky et al (Egypt September 2017- December 2018) [48]	9,4		16,7	8,7	8,7	9,4
Dilnessa and Bitew (Ethiopia 2013- 2014) [14]	20,3		11,8	0	0	
P.WAITAYANGKOON (Thailand 2017) [39]	12		15	23	20	
L. Garriga (Spain July 2017- June 2018) [95]	10,2	29,3	12,9		0	0
This study (Tunisia 2012-2022)	16,6	17,9	23,3	19,8	18,7	24,5

Pus sup: superficial pus; Pus prof: deep pus; HC: blood culture; Res: respiratory sampling; ECBU: urine cytobacteriological examination; Mat. Materials.

6. SELF-CRITICISM OF THE STUDY

It is important to mention some of the limitations of this work. On the one hand, this is a retrospective study in which we are limited by the clinical data, which may be incomplete or may not include all the relevant variables. On the other hand, given the large size of the sample and the lack of computerisation in our hospital, we cannot consult all the patients' files and therefore we will not have access to clinical and epidemiological data and we cannot have an idea of the clinical revolution of the patients. What's more, given the lack of clinical information and the high frequency of MRSA carriage, we haven't been able to find out whether the strains we are isolating are really pathogenic or colonising.

5 CONCLUSION

S, aureus is the most pathogenic species of the Staphylococcus genus. This ubiquitous bacterium can exist as a commensal bacterium of the skin and mucous membranes. However, in certain circumstances, it can become pathogenic and cause infections ranging from benign skin infections to serious conditions such as pneumonia and septicaemia. The balance between commensal and pathogenic status depends on a number of factors, including the host's immune system and ability to evade its defences or produce toxins. Staphylococcal infections are a major public health problem, both because of the virulence of this pathogen and because of the emergence of strains that are resistant to various antimicrobial agents, including methicillin.

In this context, this study focused on the epidemiology of this bacterial strain in the microbiology laboratory of the Fattouma Bourguiba University Hospital in Monastir over an 11-year period from January 2012 to December 2022. All samples from surgery, medicine, intensive care, paediatrics, emergency and health facilities of the same university hospital were included. The aim was to determine the frequency of *S. aureus* isolates in the various samples from the various departments, and to report on changes in the antibiotic resistance of these strains over the years, so that appropriate therapeutic and preventive strategies could be implemented.

During the study period, we included 4541 strains *of S, aureus*, representing a prevalence of 13%. This prevalence was relatively constant over the years, with the exception of 2022, when a peak of 19% (n=491) was observed. Most of the strains were isolated from surgical departments (n=1851, 40.8%), mainly from the orthopaedic department (n=799, 17.6%), and from medical departments, mainly from the nephrology department. The evolution of the percentages *of S. aureus* in the various departments was stable throughout the period. This germ was found in different anatomical sites in varying proportions. Approximately half of the *S, aureus* strains were isolated from superficial pus samples (n = 1916, or 42.2%), with blood cultures occupying 2eme place (n = 825, or 18.2%), followed by deep pus samples (n = 762, or 16.8%). The levels of *S. aureus* in blood cultures and deep pus samples were virtually constant. However, a decrease in the percentage of *S, aureus* strains was observed in 2017, 2020, 2021 and 2022 for superficial pus samples.

The distribution of *S. aureus* strains by sample differed from one department to another. In the surgical wards, pus samples predominated throughout the study years. Almost all respiratory samples were collected in intensive care units (72.2%, n=317) and the frequency of *S, aureus* in this type of sample increased considerably over the years, reaching 81% (n=84) in 2022. *S, aureus-positive* blood cultures were mainly observed in medical departments (52.6%, n=434). However, the trend was marked by a predominance of strains in superficial pus between 2012 and 2016, and from 2017 onwards, blood cultures took first place. A study of the resistance of these strains to different antibiotics showed high rates of resistance, mainly to penicillin G (96.4%), kanamycin (28%), erythromycin (13.8%), tetracyclines (28.9%), fusidic acid (27.1%) and fluoroquinolones. Other antibiotics with low levels of resistance to this bacterial strain were glycopeptides and chloramphenicol. Antibiotic resistance was relatively stable during the study period for all therapeutic classes with the exception of glycopeptides and fusidic acid. In fact, there was a constant increase in the rate of resistance of strains to fusidic acid, and resistance to glycopeptides became considerably higher from 2020 onwards.

The highest levels of resistance were observed, mainly in intensive care and medical wards, in the sampling of materials, punctures and ECBU.

Methicillin resistance is a major cause of morbidity and mortality worldwide. During the study period, 828 MRSA strains (18.6%) were isolated. This rate was relatively stable over the study period. Unlike *S. aureus* strains as a whole, these strains predominated in intensive care units (28.8%, n=148), the emergency department (20.3%, n=30) and surgical units (18.8%, n=341), mainly in the general surgery unit (24.4%, n=76); mainly in blood cultures (23.3%, n=186) and materials (24.5%, n=34).

Certain strains of MRSA have become resistant to virtually all antibiotics. In addition, greater resistance was noted to kanamycin (63.1%), erythromycin (27%), fluoroquinolones, tetracycline (51.5%) and fusidic acid (43.6%). Changes in the MRSA rate varied greatly depending on the department.

The antimicrobial agents that remain active on methicillin-resistant strains are glycopeptides, making these antibiotics the reference treatment. Nevertheless, an increase in resistance to this family of antibiotics has been described. In view of the depletion of therapeutic resources, preventive measures must be applied - to avoid the emergence of MRSA strains.

-Recommendations and outlook :

In order to optimise microbiological diagnosis, it is recommended that all prescriptions be accompanied by the clinical information needed to interpret the results correctly. In addition, swab samples should be avoided in favour of puncture or aspiration using a syringe, in order to differentiate between true infection and colonisation.

The increase in antibiotic resistance in staphylococcus aureus is a major cause for concern. In order to limit the rate of MRSA isolation, it is recommended to collaborate with clinical colleagues, particularly infectious diseases specialists, in order to set up a decolonisation protocol or treatment adapted to the Fattouma Bourguiba University Hospital in Monastir. A referent in antibiotic therapy should also always be consulted to ensure appropriate use of antibiotics and minimise the risk of antimicrobial resistance. In addition, we need to implement a policy of screening patients for reservoirs of Multi-Resistant Bacteria (MRB) in all hospital departments. In our hospital, screening is only carried out in intensive care units.

In order to limit the rate of isolation of MRSA strains, standard precautions must be followed, in particular:

- Hand washing: Practise rigorous hand hygiene with antibacterial soap
- Appropriate use of antibiotics: Avoiding overuse or inappropriate use of antibiotics to prevent bacterial resistance
- Isolating infected patients: Implement isolation measures for patients carrying MRSA to prevent transmission.
- Cleaning and disinfection: Ensuring frequent disinfection of potentially contaminated surfaces and objects.
- Wearing gloves and masks: Use personal protective equipment to prevent transmission.
- Education and awareness: informing medical staff, patients and the public about preventive practices.
- Surveillance and control: Set up surveillance programmes to detect MRSA cases quickly and take preventive measures.

It is essential to combine these measures to minimise the spread of MRSA and preserve the effectiveness of antibiotic treatment. This is one of the CLIAS (Comite de Lutte contre les Infections Associees aux Soins) priorities, alongside gram-negative bacilli.

<u>Our prospects are :</u>

- Having rapid MRSA screening facilities
- Advanced molecular studies to speed up identification of *S. aureus* and detect MRSA
- Developing molecular technologies (Nucleic Acid Amplification Tests or NAATs)
- Improving the identification and study of antibiotic resistance in *S. aureus* strains
- Digging into the research trail to detect virulence and resistance genes.

6 BIBLIOGRAPHICAL REFERENCES

1. Gherardi G. *Staphylococcus aureus* Infection: Pathogenesis and Antimicrobial Resistance. *Int J Mol Sci 2023;24:8182.*

2. Paling FP, Hazard D, Bonten MJ, Goossens H, Jafri HS, Malhotra-Kumar S, et al. Association of *Staphylococcus aureus* colonization and pneumonia in the intensive care unit. *JAMA Netw Open 2020;3:e2012741.*

3. Cheung GY, Bae JS, Otto M. Pathogenicity and virulence of *Staphylococcus aureus*. *Virulence 2021;12:547-69.*

4. Ahmad-Mansour N, Loubet P, Pouget C, Dunyach-Remy C, Sotto A, Lavigne JP, et al. *Staphylococcus aureus* toxins: an update on their pathogenic properties and potential treatments. *Toxins 2021;13:677.*

5. Aratani T, Tsukamoto H, Higashi T, Kodawara T, Yano R, Hida Y, et al. Association of methicillin resistance with mortality of hospital-acquired *Staphylococcus aureus* bacteremia. *J Int Med Res 2021;49(11):1-12.*

6. Dujardin M, Beti E, Vanhecke C. Pneumopathie necrosante a *Staphylococcus aureus*. *Ann Fr Med Urgence 2018;8:178-80.*

7. Tong SY, Davis JS, Eichenberger E, Holland TL, Fowler VG Jr. *Staphylococcus aureus* infections: epidemiology, pathophysiology, clinical manifestations, and management. *Clin Microbiol Rev 2015;28:603-61.*

8. Wu Q, Sabokroo N, Wang Y, Hashemian M, Karamollahi S, Kouhsari E. Systematic review and meta-analysis of the epidemiology of vancomycin-resistance *Staphylococcus aureus* isolates. *Antimicrob Resist Infect Control 2021;10:101.*

9. Abubakar U, Sulaiman SAS. Prevalence, trend and antimicrobial susceptibility of Methicillin Resistant *Staphylococcus aureus* in Nigeria: a systematic review. *J Infect Public Health 2018;11:763-70.*

10. Koinam FR, Guira F, Somda NS, Yameogo A, Bonkoungou IJ, Traore Y, et al. Profile of sensitivity and resistance to antibiotics of *Staphylococcus aureus* strains isolated from patients fluids in medical biology department of national public health laboratory of Ouagadougou, Burkina Faso. *J Fundam Appl Sci 2016;9:553-66.*

11. Al Musawi S, Alkhaleefa Q, Alnassri S, Alamri AM, Alnimr A. Eleven-year surveillance of methicillin-resistant *Staphylococcus aureus* infections at an Academic Health Centre. *J Prev Med Hyg 2022;63:E132-8.*

12. Canadian Nosocomial Infection Surveillance Program. Healthcare-associated infections and antimicrobial resistance in Canadian acute care hospitals, 2017 to 2021. *Canadian Communicable Disease Report 2023;49:263-81.*

13. Chelkeba L, Melaku T. Epidemiology of staphylococci species and their antimicrobial-resistance among patients with wound infection in Ethiopia: a systematic review and meta-analysis. *J Global Antimicrob Resist 2022;29:483-98.*

14. Dilnessa T, Bitew A. Prevalence and antimicrobial susceptibility pattern of methicillin resistant *Staphylococcus aureus* isolated from clinical samples at Yekatit 12 Hospital Medical College, Addis Ababa, Ethiopia. *BMC Infect Dis 2016;16:398.*

15. Shahi K, Raj Rijal K, Adhikari N, Upendra Thapa S, Raj Banjara M, Sharma VK, et al. Methicillin resistant *Staphylococcus aureus*: prevalence and antibiogram in various clinical specimens at Alka Hospital. *Tribhuvan Univ J Microbiol 2018;5:77-82.*

16. Noel G, Benigne B, Vincent M. Prevalence of *Staphylococcus aureus* among clinical isolates and their responses to selected antibiotics at Centre Hospitalier Universitaire de Kigali (CHUK). *J Microbiol Exp 2017;5:11-12.*

17. Idrees MM, Saeed K, Shahid MA, Akhtar M, Qammar K, Hassan J, et al. Prevalence of mecA- and mecC-Associated Methicillin-Resistant *Staphylococcus aureus* in Clinical Specimens, Punjab, Pakistan. *Biomedicines 2023;11:878.*

18. Sapkota J, Sharma M, Jha B, Bhatt CP. Prevalence of *Staphylococcus aureus* isolated from clinical samples in a tertiary care hospital: a descriptive cross-sectional study. *J Nepal Med Assoc 2019;57(220):398-402.*

19. Lahlou YB, Comlan AG, Maleb A, Lemnouer A, Elouennass M. Epidemiology and resistance profile of Gram-positive bacteria at the Mohammed V military training hospital in Rabat, Morocco. *J Mar Sci Med 2020;22;2:20-4.*

20. European Centre for Disease Prevention and Control. Antimicrobial resistance in the EU/EEA (EARS-NET) Annual Epidemiological Report for 2019. *Stockholm: ECDC; 2020.*

21. European Centre for Disease Prevention and Control. Antimicrobial resistance in the EU/EEA (EARS-Net) Annual Epidemiological Report 2021. *Stockholm: ECDC; 2022.*

22. Kika B, Abazaj E, Petri O, Koraqi A. Prevalence of *Staphylococcus aureus* and methicillin-resistant *S. aureus* in adult hospitalized patients in "Mother Theresa" Hospital Center in Albania. *Open Access Maced J Med Sci 2020; 8:360-4.*

23. European Centre for Disease Prevention and Control. Antimicrobial resistance in the EU/EEA (EARS-NET) Annual Epidemiological Report for 2020. *Stockholm: ECDC; 2022.*

24. Mendes RE, Sader HS, Castanheira M, Flamm RK. Distribution of main

Gram-positive pathogens causing bloodstream infections in United States and European hospitals during the SENTRY Antimicrobial Surveillance Program (2010-2016): concomitant analysis of oritavancin in vitro activity. *J Chemother 2018;30:280-9.*

25. Frikh M, Lemouer A, Belfquih B. Sensitivity profiles of *Staphylococcus aureus* isolates Retrospective study over Eight years in a teaching hospital. *J Mar Sci Medicales 2016;20:5-9.*

26. Boutiba Ben Boubaker I, Saidani M. Antibio-Resistance in Tunisia. LART: Data 2017: *S. aureus* (N=1790). *Tunis: LART; 2018.*

27. Boutiba Ben Boubaker I, Saidani M. Antibio-Resistance in Tunisia LART: Data 2012. *S. aureus* (N=2102). *Tunis: LART; 2013.*

28. Boutiba Ben Boubaker I, Saidani M. Antibio-Resistance in Tunisia LART: Data 2015. *S. aureus* (N=1757). *Tunis: LART; 2016.*

29. Boutiba Ben Boubaker I, Saidani M. Antibio-Resistance in Tunisia LART: Data 2016. *S. aureus* (N=1496). *Tunis: LART; 2017.*

30. Miamina FA, Mtapuri-Zinyowera S, Najjuka G, Perovic O, Zayed B, Ahmed YA, Ismail MT, Pessoa da Silva CL. Global Antimicrobial Resistance and Use Surveillance System on the African continent: Early implementation 2017-2019. *Afr J Lab Med 2022;11:1594.*

31. Migaud H, Senneville E, Gougeon F, Marchetti E, Amzallag M, Laffargue P. Risque infectieux en chirurgie orthopedique. *EMC- Techniques Chirurgicales - Orthopedie - Traumatologie 2005:1-15 [Article 44-005].*

32. Elifranji ZO, Haddad B, Salameh A, Alzubaidi S, Yousef N, Al Nawaiseh M, et al. Microbiological profile and drug resistance analysis of postoperative infections following orthopedic surgery: a 5-year retrospective review. *Adv Orthop 2022;2022:7648014.*

33. Otto M. Staphylococcal biofilms. *Microbiol Spectr 2018;6(4):1-26*

34. Direction de la Pharmacie et du Medicament. Antibiotic therapy for bone and joint infections iiiguc's infections caused by pyogenes: national recommendations. *[Online]. 2006 [Accessed 21/05/2024], available at URL:https://www.infectiologie.org.tn/pdf_ppt_docs/ recommandations/1536342830. pdf*

35. Gangoue LG, Nguimbi E, Ondzia FRO, Mahoukou BPB, Lingouagou TM, Ngoulou TB, et al. Prevalence, resistance to quinolones/ fluoroquinolones of *Staphylococcus aureus* strains isolated in the hospital and university center of Brazzaville, Republic of Congo. *Adv Microbiol 2021;11:607-15.*

36. Nabti I., Tieb C. Les infections a *Staphylococcus aureus* au CHU de Constantine [Master's thesis]. *Constantine: Universite Freres Mentouri, Faculte des Sciences de la Nature et de la Vie; 2019.*

37. ONERBA. Activity report 2017. *[Online]. 2018 [Accessed 21/05/2024], available at URL: http://onerba.org/publications/rapports- onerba/?cn-reloaded=1*

38. Worku M, Belay G, Tigabu A. Bacterial profile and antimicrobial susceptibility patterns in cancer patients. *PLoS One 2022;17:e0266919.*

39. Waitayangkoon P, Thongkam A, Benjamungkalarak T, Rachayon M, Thongthaisin A, Chatsuwan T, et al. Hospital epidemiology and antimicrobial susceptibility of isolated methicillin-resistant *Staphylococcus aureus*: a one-year retrospective study at a tertiary care center in Thailand. *Pathog Glob Health 2020;114:212-7.*

40. Kanjilal S, Sater MR, Thayer M, Lagoudas GK, Kim S, Blainey PC, Grad YH. Trends in Antibiotic Susceptibility in *Staphylococcus aureus* in Boston, Massachusetts, from 2000 to 2014. *J Clin Microbiol 2017;56:e01160-17.*

41. Alshomrani MK, Alharbi AA, Alshehri AA, Arshad M, Dolgum S. Isolation of *Staphylococcus aureus* Urinary Tract Infections at a CommunityBased Healthcare Center in Riyadh. *Cureus 2023;15:e35140.*

42. Societe Frangaise de Microbiologie. EUCAST/CA-SFM Recommandations 2022 V.1.0 *[Online]. 2022 [Accessed 26/05/2024], available at URL: https://www.sfm-microbiologie.org/wp-content/ uploads/2022/05/CASFM2022_V1.0. pdf*

43. Romano-Bertrand S, Pozzetto B, Decousser JW. Antibiotic susceptibility testing: objectives, methods and interpretation. *Hygienes 2023;31:253-8.*

44. Naimi HM, Rasekh H, Noori AZ, Bahaduri MA. Determination of antimicrobial susceptibility patterns in *Staphylococcus aureus* strains recovered from patients at two main health facilities in Kabul, Afghanistan. *BMC Infect Dis 2017;17:737.*

45. Saleem F, Fasih N, Zafar A. Susceptibility pattern of methicillin resistant *Staphylococcus aureus* to vancomycin and other alternate agents: report from a private sector hospital laboratory. *J Pak Med Assoc 2017;67:1743-6.*

46. Merah I, Bouguerra M, Moussaoui K, Merani N. Mecanisme moleculaire de la résistance bacterienne aux antibiotiques cas de *Staphylococcus aureus* [Memoire de Master]. Guelma: Universite 8 Mai 1945 Faculte des Sciences de la Nature et de la *Vie et Sciences de la Terre et de l'Univers; 2021.*

47. Tara T, Khan Z, Ali H, Sardar S, Samad A, Ullah K, et al. A review of global epidemiology and antibiotic resistance of *Staphylococcus aureus. Pak J Med Health Sci 2021;15:3921.*

48. Alfeky AE, Tawfick MM, Ashour MS, El-Moghazy AA. High prevalence of multi-drug resistant methicillin-resistant *Staphylococcus aureus* in tertiary

Egyptian hospitals. *J Infect Dev Ctries 2022;16:795-806.*

49. Coombs GW, Daley DA, Yee NW, Shoby P, Mowlaboccus S. Australian Group on Antimicrobial Resistance (AGAR) Australian *Staphylococcus aureus* Sepsis Outcome Programme (ASSOP) Annual Report 2020. *Commun Dis Intell 2022;46.*

50. McNeil JC, Sommer LM, Vallejo JG, Boyle M, Hulten KG, Kaplan SL, et al. Going back in time: increasing penicillin susceptibility among methicillin-susceptible Staphylococcus aureus osteoarticular infections in children. *Antimicrob Agents Chemother 2023;67:e0119622.*

51. Tasneem U, Majid M, Mehmood K, Redaina, Rehman FU, Andleeb S, et al. Co-occurrence of antibiotic resistance and virulence genes in Methicillin Resistant *Staphylococcus aureus* (MRSA) Isolates from Pakistan. *Afri Health Sci 2022;22:486-95.*

52. Raut S, Bajracharya K, Adhikari J, Pant SS, Adhikari B. Prevalence of methicillin resistant *Staphylococcus aureus* in Lumbini medical college and teaching hospital, Palpa, Western Nepal. *BMC Res Notes 2017;10:187.*

53. Qodrati M, SeyedAlinaghi S, Dehghan Manshadi SA, Abdollahi A, Dadras O. Antimicrobial susceptibility testing of *Staphylococcus aureus* isolates from patients at a tertiary hospital in Tehran, Iran, 2018-2019. *Eur J Med Res 2022;27:152.*

54. Khanal A, G C S, Gaire A, Khanal A, Estrada R, Ghimire R, Panthee S. Methicillin-resistant *Staphylococcus aureus* in Nepal: A systematic review and meta-analysis. *Int J Infect Dis. 2021 Feb;103:48-55.*

55. Talapan D, Sandu AM, Rafila A. Antimicrobial Resistance of *Staphylococcus aureus* Isolated between 2017 and 2022 from Infections at a Tertiary Care Hospital in Romania. *Antibiotics 2023;12:974.*

56. Carrel M, Smith M, Shi Q, Hasegawa S, Clore GS, Perencevich EN, et al. Antimicrobial resistance patterns of outpatient *Staphylococcus aureus* isolates. *JAMA Netw Open 2024;7:e2417199.*

57. La Vecchia A, Ippolito G, Taccani V, Gatti E, Bono P, Bettocchi S, et al. Epidemiology and antimicrobial susceptibility of Staphylococcus aureus in children in a tertiary care pediatric hospital in Milan, Italy, 2017-2021. *Ital J Pediatr 2022;48:67.*

58. Klos M, Pomorska-Wesolowska M, Romaniszyn D, Chmielarczyk A, Wojkowska-Mach J. Epidemiology, drug resistance, and virulence of *Staphylococcus aureus* isolated from ocular infections in Polish patients. *Pol J Microbiol 2019;68:541-8.*

59. Singh K, Raju V, Nikalji R, Jawale S, Patel H, Ahdal J, et al. Methicillin- resistant *Staphylococcus aureus* infections in patients with renal

disorders: a review. *World J Nephrol Urol 2019;8:8-13.*
60. Szymanek-Majchrzak K, Mlynarczyk A, Kawecki D, Pacholczyk M, Durlik M, Deborska-Materkowska D, et al. Resistance to aminoglycosides of methicillin-resistant strains of *Staphylococcus aureus*, originating in the surgical and transplantation wards of the Warsaw clinical center - a retrospective analysis. *Transplant Proc 2018;50:2170-5.*
61. Alle Akakpo AE. Meta analyse de la sensibilite aux antibiotiques des bacteries Isolees dans les différents produits pathologiques au mali de 2000 a 2020 [These]. *Bamako: Universite des Sciences, des Techniques et Technologies, Faculte de Medecine et d'Odonto-Stomatologie; 2021.*
62. ONERBA. Activity report 2018. *[Online]. 2019 [Accessed 21/05/2024], available at URL: http://onerba.org/publications/rapports- onerba/*
63. Dikoumba A, Onanga R, Nguema P, Mangouka L, Iroungou B, Kassa F, et al. Phenotipic prevalence of antibiotic resistance in Gabon. *Open J Med Microbiol 2020;11:100-18.*
64. Ade TI, Osiyemi JA, Aso RE, Akinduti PA, Sunmola NO. Prevalence of macrolide-lincosamide-streptogramin-B resistance among clinical *Staphylococcus aureus* isolates in university of Ilorin teaching hospital, Ilorin, Nigeria. *Afr J Clin Exp Microbiol 2022;23:168-73.*
65. Khodabandeh M, Mohammadi M, Abdolsalehi MR, Alvandimanesh A, Gholami M, Bibalan MH, et al. Analysis of Resistance to Macrolide-Lincosamide-Streptogramin B Among mecA-Positive *Staphylococcus aureus* Isolates. *Osong Public Health Res Perspect 2019;10:25-31.*
66. Shariati A, Dadashi M, Chegini Z, van Belkum A, Mirzaii M, Khoramrooz SS, et al. The global prevalence of Daptomycin, Tigecycline, Quinupristin/Dalfopristin, and Linezolid-resistant *Staphylococcus aureus* and coagulase-negative staphylococci strains: a systematic review and metaanalysis. *Antimicrob Resist Infect Control 2020;9:56.*
67. Quincampoix JC, Mainardi JL. Mechanisms of resistance in gram-positive cocci. *Reanimation 2001;10:267-75.*
68. Haute Autorite de Sante. Prise en charge des infections cutanees bacteriennes courantes: Methode Recommandations pour la pratique clinique. *Paris: HAS; 2019.*
69. Mezghani Maalej S, Malbruny B, Leclercq R, Hammami A. Emergence of *Staphylococcus aureus* strains resistant to pristinamycin in Sfax (Tunisia). *Pathol Biol (Paris) 2012;60(6):e71-4.*
70. Verneuil L, Marchand C, Vidal JS, Ze Bekolo R, Daurel C, Lebouvier G, et al. Factors associated with emergence of pristinamycin-resistant *Staphylococcus aureus* in a dermatology department: a case-control study. *Br J*

Dermatol 2010;163:329-33.
71. Gade ND, Qazi MS. Fluoroquinolone therapy in *Staphylococcus aureus* infections: where do we stand? *J Lab Physicians 2013;5:109-12.*
72. de Lastours V, Fantin B. Resistance to fluoroquinolones in 2010: what impact for prescribing in intensive care. *Reanimation 2010;19:347-53.*
73. Haute Autorite de Sante. Strategie d'antibiotherapie et prevention des resistances bacteriennes en etablissement de sante. Recommandation de bonne pratique. *Paris: HAS; 2008.*
74. Xu Y, Wang B, Zhao H, Wang X, Rao L, Ai W, et al. In vitro activity of vancomycin, teicoplanin, linezolid and daptomycin against methicillin- resistant *Staphylococcus aureus* isolates collected from Chinese hospitals in 2018-2020. *Infect Drug Resist 2021;14:5449-56.*
75. Wilcox M, Al-Obeid S, Gales A, Kozlov R, Martmez-Orozco JA, Rossi F, et al. Reporting elevated vancomycin minimum inhibitory concentration in methicillin-resistant *Staphylococcus aureus*: consensus by an International Working Group. *Future Microbiol 2019;14:345-52.*
76. Gourdeau M, Le Guerrier P, Pilon PA, St-Antoine P, Dolce P, Trudeau G. Les Infections Nosocomiales. Mesures de contrôle et de prévention des infections a Staphylococcus aureus ayant une sensibilité reduite a la vancomycine au Quebec: Lignes directrices provisoires pour les centres hospitaliers quebecois. *Quebec: Sante et Services Sociaux; 2002.*
77. Lee YC, Chen PY, Wang JT, Chang SC. Prevalence of fosfomycin resistance and gene mutations in clinical isolates of methicillin-resistant *Staphylococcus aureus*. *Antimicrob Resist Infect Control 2020;9:135.*
78. Chen T, Zhao L, Liu Y, Wang Y, Jian Y, Zhao N, et al. Mechanisms of high-level fosfomycin resistance in *Staphylococcus aureus* epidemic lineage ST5. *JAntimicrob Chemother 2022;77:2816-26.*
79. Amissah NA, van Dam L, Ablordey A, Ampomah OW, Prah I, Tetteh CS, et al. Epidemiology of Staphylococcus aureus in a burn unit of a tertiary care centre in Ghana. *PLoS One 2017;12:e0181072.*
80. Hajikhani B, Goudarzi M, Kakavandi S, Amini S, Zamani S, van Belkum A, et al. The global prevalence of fusidic acid resistance in clinical isolates of *Staphylococcus aureus*: a systematic review and meta-analysis. *Antimicrob Resist Infect Control 2021;10:75.*
81. Haute Autorite de Sante. Commission de la transparence synthese d'avis 13 octobre 2021: Acide fusidique. *Paris: HAS; 2021.*
82. Hull RC, Wright RCT, Sayers JR, Sutton JAF, Rzaska J, Foster SJ, et al. Antibiotics limit adaptation of drug-resistant *Staphylococcus aureus* to hypoxia. *Antimicrob Agents Chemother 2022;66:e0092622.*

83. Kriegeskorte A, Lore NI, Bragonzi A, Riva C, Kelkenberg M, Becker K, et al. Thymidine-Dependent *Staphylococcus aureus* small-colony variants are induced by trimethoprim-sulfamethoxazole (SXT) and have increased fitness during SXT challenge. *Antimicrob Agents Chemother 2015;59:7265-72.*
84. Wang C, Fang R, Zhou B, Tian X, Zhang X, Zheng X, et al. Evolution of resistance mechanisms and biological characteristics of rifampicin-resistant *Staphylococcus aureus* strains selected in vitro. *BMC Microbiol 2019;19:220.*
85. Huang J, Zhang F, Zhang J, Dai J, Rong D, Zhao M, et al. Molecular Characterization of rifampicin-resistant *Staphylococcus aureus* isolates from retail foods in China. *Antibiotics (Basel) 2021;10:1487.*
86. Vestergaard M, Frees D, Ingmer H. Antibiotic resistance and the MRSA Problem. *Microbiol Spectr 2019;7(2).*
87. Ahoyo AT, Baba-Moussa L, Makoutode M, Gbohoun A, Bossou R, Dramane K, et al. Incidence of meticillin-resistant *Staphylococcus aureus* in the neonatology department of the Zou et Collines departmental hospital in Benin. *Arch Pediatr 2006;13:1391-6.*
88. Zerouki A, Abada S, Tali-Maamar H, Rahal K, Naim M. Characterisation of meticillin-resistant *Staphylococcus aureus* surgical site infections in orthopaedic and trauma surgery in an Algerian hospital. *Rev Chir Orthop Traumatol 2015;101:176-80.*
89. Friedman L; Wallar L; Papadopoulos A. Environmental risk factors for community-acquired MRSA. *Vancouver, BC: National Collaborating Centre for Environmental Health; 2015.*
90. Kouatchet A, Eveillard M. Meticillin-resistant *Staphylococcus aureus* in intensive care. *Med Intensive Reanim 2012;21:493-505.*
91. Massaly A, Ka D, Shall K, Badiane AS, Fall NM, Diallo Mbaye K, et al. Profile of healthcare-associated infections and antibiotic resistance in an infectious diseases department (Dakar, Senegal). *Jaccr Infectiology 2022;4(2):1-10.*
92. Jamard S, Ferry T, Valour F. Physiopathology of osteoarticular infections. *Rev Rhum Monogr 2022;89:3-10.*
93. Kimmig A, Hagel S, Weis S, Bahrs C, Loffler B, Pletz MW. Management of *Staphylococcus aureus* Bloodstream Infections. *Front Med (Lausanne) 2021;7:616524.*
94. Mesrati I, Saidani M, Jemili M, Ferjeni S, Slim A, Boubaker IB. Virulence determinants, biofilm production and antimicrobial susceptibility in *Staphylococcus aureus* causing device-associated infections in a Tunisian hospital. *Int J Antimicrob Agents 2018;52:922-9.*
95. Garriga Ferrer-Bergua L, Borrull Senra AM, Perez Velasco C, Montero

Valladares C, Collazo Vallduriola I, Moya Villanueva S, et al. Rate of methicillin-resistant *Staphylococcus aureus* in pediatric emergency departments in Spain *An Pediatr (Barc) 2022;97:95-102.*

MIX
Papier aus verantwortungsvollen Quellen
Paper from responsible sources
FSC® C105338

Printed by Books on Demand GmbH, Norderstedt / Germany